WESTCHESTER PUBLIC
W9-BTD-736

For Mom—who has guided our family through
more than fifty years of fruit-filled seasons and
with whom I first learned to bake.

SIMPLE FRUIT

Seasonal Recipes for Baking, Poaching, Sautéing, and Roasting

LAURIE PFALZER

PHOTOGRAPHY BY CHARITY BURGGRAAF

CONTENTS

Kitchen Memories vii

Introduction viii

SPRING 1

SUMMER 34

AUTUMN 121

WINTER 157

Acknowledgments 189

Index 191

RECIPE LIST

Spring

RHUBARB 3

Vanilla Roasted Rhubarb 5

Rhubarb-Ginger Sorbet 6

Straight Rhubarb Pie 9

Rhubarb Fool 14

STRAWBERRIES 17

Strawberry Pavlova 19

Strawberry-Lime Layer Cake 23

Strawberry-Coconut Sorbet 28

Roasted Strawberry Ice Cream 32

Summer

CHERRIES 37

Cherry Hand Pies 39

Smoky Sweet Cherries in Port with Bittersweet Chocolate 41

Sour Cherry Compote 45

White Chocolate Mousse with Poached Sweet Cherries 46

RASPBERRIES 51

Raspberry and Lemon Balm Shortcake 52

Port Sabayon with Fresh Raspberries 55

Raspberry Custard Tart 56

Dark Chocolate Cake with Raspberry-Orange Compote 62

BLUEBERRIES 65

Classic Lattice Blueberry Pie 67

Blueberry-Cinnamon Crepes 70

Blueberry and Lemon Curd Tart 76

Poached Blueberries with Vanilla Bean and Anise 79

BLACKBERRIES AND MARIONBERRIES 81

Marionberry Crostata with Whole Grain Crust 82

Pain Perdu with Assorted Berries and Grand Marnier 86

Bittersweet Chocolate Tart with Blackberries and Basil 89

Vanilla Bean Cake with Glazed Blackberries and Stone Fruits 91

APRICOTS 97

Grilled Apricots with Brown Butter and Maple-Tamari Glaze 98

Apricot and Walnut Streusel Bars 102

Poached Apricots in Sauternes 105

PEACHES AND NECTARINES 109

Peaches and Nectarines Sautéed with Thyme 110

Peach Mousse 114

Roasted Peach Bread Pudding 116

Autumn

PLUMS 123

Simple Stewed Plums and Pluots 124

Plum-Lavender Crisp 127

Fromage Blanc Cheesecake with Brown-Butter-Braised Plums 128

APPLES 133

Rosemary Apples in Crepes with Rum Caramel Sauce 134

Brandied Apple Pie with Ginger Streusel 138

Apple Upside-Down Cake with Spelt and Rye Flours 142

Cider-Braised Apples with Warm Winter Spices 144

PEARS 147

Pear and Fig Pie 148

Burgundy-Poached Pears with Mulled-Wine Syrup 152

Maple and Pear Panna Cotta 154

Winter

CRANBERRIES 159

Mom's Cranberry-Walnut Pie 160

Apple-Cranberry Tarte Tatin 164

Cranberry and Bay Upside-Down Cake 166

CITRUS 169

Meyer Lemon Buttermilk Sherbet 170

Almond Cake with Warm Citrus and Thyme 173

Orange and Rye Madeleines 175

Orange, Rosemary, and Hazelnut Brittle 177

DRIED FRUITS 181

Grandma's Dark Holiday Fruitcake 182

Apricot-Almond Scones 186

Kerr

KITCHEN MEMORIES

The summer I wrote this book, my parents were preparing to move out of our family home of fifty years. It was the only home they'd owned and the only home my sisters and brothers and I remember as children. During that summer, we prepared for the big move, going through drawers and closets with old photos, games, books, and mementos. I crawled deep under the staircase, where my mother stored her canned peaches, pears, plums, and pickles. And we scoured cupboards and drawers in the kitchen, where every summer we had washed, peeled, cored, cut, froze, canned, jellied, and jammed virtually every local fruit. We perused Mom's cookbooks (and her mother's cookbooks) with the family recipes she had made for us for over fifty years. We looked out the windows of the kitchen to the backyard, listened to music on my dad's old stereo, and reminisced about the many meals we had enjoyed in that kitchen. The last thing my younger sister and I did at the house was pick the final crop of Dad's Concord grapes, which ripen in October. Clipping off each individual bunch, which hung heavy on the arbor that shaded the porch swing, we took time to swing where my dad had swung three generations of our family. It was his favorite place to sit, relax, and ponder.

A few weeks before the house closed, I went back for one last look. Despite some improvements, it still felt like our house. As I walked through each room, I could hear our voices. Memories washed over me. When I reached the kitchen, which looked out over the backyard, I stood at the sink and felt the comfort and closeness of that room. The apparitions of fifty years of fruitful seasons poured from the walls. Suddenly, boxes of apples from my grandmother's Transparent tree appeared around me. I saw my mom at the stove dropping canning jars of peaches into the steaming water. My sister was next to me, hulling strawberries for jam and freezing. Looking out the window, my dad stood on the ladder at the Concord grape arbor over the swing, picking the grapes. At the other end of the yard, stiff stalks of rhubarb grew next to the fence. Everything had changed that summer, and yet nothing had changed. I stood in the kitchen and sobbed, and then I left the house for the last time.

INTRODUCTION

My mom never put spices in her apple pie. Some might think this strange—no cinnamon in an apple pie? But her pies were legendary, made with her perfectly handled pie crust and Transparent apples from my grandmother's tree in south Seattle. The lucky recipients of her pies never missed the cinnamon. It was from her that I first learned to respect the integrity of the fruit, not overspicing or covering up the flavor of the local fruit we had just picked. Whether we had rhubarb from our backyard garden, Italian prunes from our tree in the backyard, or strawberries from the local patches, my mom let the fruit shine.

Baking has always been a large part of my life because I grew up in a home with fabulous homemade pies, cakes, and other baked goods and desserts. My mom always made it look easy, and I grew up baking at her side without any fears about making a mistake. It wasn't until I began to teach home cooks that I realized the challenges my students faced, not having grown up with a baking influence and trying to learn.

I didn't go the professional pastry route until my midthirties, and I feel this path has made all the difference to teaching home cooks. This unique view of baking and pastry—learning as a home cook and then a professional cook—is what I bring to my students and to this book. You will find plenty of tips I learned in professional kitchens—like Salish Lodge and Spa and King Arthur Flour Bakery—which I have adapted for the home cook, as well as tips I've learned through ten years of teaching. My goal is to lead home cooks through these recipes with inspiration, vitality, and encouragement. I often see students worrying too much about whether they'll do it right rather than just getting in there and doing it. I want to help you feel confident and at ease baking and cooking with fruit, just as I try to do with my students.

In this book, you will find an array of fruit recipes, each utilizing a particular cooking technique, along with a handful of recipes using fresh fruit. I've carefully paired the fruits with complementary flavors in the form of an herb, a spice, a liqueur, or possibly another fruit. Where appropriate, I offer suggestions for different flavor pairings or a substitute fruit. Each recipe enhances the inherent flavor of the fruit, most often through a cooking technique—poaching, roasting, braising, baking, or sautéing—which transforms the fruit. There are many simple recipes in this book, including fruit desserts and other preparations that could serve as desserts on their own or as accompaniments for cake, ice cream, or your morning granola. A small group of recipes constructs more elaborate desserts to be savored for a special occasion or after a wonderful meal. And plenty of recipes can be used with many different fruits, so you will find these recipes multiseasonal and easily adaptable.

INGREDIENTS AND EQUIPMENT

Ingredients

Eggs and dairy: Baking is a game of consistency, so try to stick with the same butter, eggs, and dairy products—such as milk and heavy cream—that are consistent in quality, a good value, and always available.

Flour: In this book, I reference a few flour types, including the usual all-purpose flour. I have noted where white pastry flour may give you a more tender pastry or how using a whole grain flour enhances flavor and texture. The bottom line on flour in any baked good is that you will get your best result (tender cake or chewy bagels) by using the right flour for the right product. Before changing up flours, consider the protein content (which leads to gluten structure) so that you achieve the desired texture.

Fresh versus frozen fruit: For many recipes, frozen fruit will work as well as fresh. If you are baking, stewing, saucing, or making compote, frozen fruit can usually replace fresh. It's best not to poach or braise previously frozen fruit, because freezing often changes the texture, especially for rhubarb, strawberries, apples, and pears. Pies and crisps are also easily made with frozen fruit.

Sugar: Although there is still a place for white sugar, we're fortunate to have a variety of sweeteners available that bring distinct flavor to fruit. Baking and cooking with fruit is the perfect place to change up sweeteners or combine them for enhanced flavor. I depend on unbleached granulated sugar, as well as light and dark brown sugars, demerara (or turbinado or muscovado), honey, and maple syrup to bring complex flavor to fruit. However, keep in mind that using these less refined sugars, such as organic granulated sugar, for cooking fruit compromises a few things in baking: the grains are generally larger, and the color result is darker. In most baking, this is a minor issue. But when you're making a white Italian butter cream or snowy meringues, only the more refined white sugar will do.

Vanilla beans: For whole vanilla beans, try to purchase them where they are sold in bulk. (You'll pay a premium for a glass vial of one or two beans in your local grocery store.) I use whole bean whenever possible, but I also consider the value of the bean in each recipe. In this book, the gentle cooking techniques of sautéing, poaching, roasting, stewing, and braising allow the whole-vanilla-bean flavor to linger and contribute to the dish. A second option is ground vanilla bean. (I use Wilderness Poets brand, available at some grocers and online.) A third option, vanilla paste, is also available from some grocers and online. When using vanilla beans, split the bean from end to end, leaving one end connected. Pry open the bean halves at the connected end and use the dull side of your knife to scrape out the pulp. Some recipes call for the pulp and bean, while others, just the pulp.

Equipment

Bench knife: This simple bread tool has become more popular in the kitchen for tasks other than dividing bread dough. Use it to scrape down your counter after working with dough or chocolate, to cut butter into small pieces for pie dough, or just for general cleaning up.

Bowl scraper: This plastic tool uses leverage when scraping or lifting stiff dough out of the bowl. If you've ever broken a spatula handle while stirring stiff dough, you know what I'm talking about. There are flexible and stiff styles, and they are inexpensive so it's worth having one of each on hand.

Microplane zester: This is the tool I see misused the most often in my classes. Check out online videos by Microplane to see it used correctly for zesting citrus fruit—which is mostly how you'll be using it in this book.

Pastry wheel: You might know this as a pizza wheel, but I recommend purchasing an actual pastry wheel, which has a stiffer blade that won't wobble (Ateco is a good brand.) You might want to have a fluted-edge pastry wheel also.

Spatulas and whisks: Spatulas and whisks are used so often, it's worth having three of each. Then you always have a clean one to grab. Be sure to use the heat-resistant silicone spatulas (with a heat-resistant handle).

SPRING

RHUBARB 3

Vanilla Roasted Rhubarb 5

Rhubarb-Ginger Sorbet 6

Straight Rhubarb Pie 9

Rhubarb Fool 14

STRAWBERRIES 17

Strawberry Pavlova 19

Strawberry-Lime Layer Cake 23

Strawberry-Coconut Sorbet 28

Roasted Strawberry Ice Cream 32

RHUBARB

As one of the first signs of spring in the Pacific Northwest, rhubarb is in good company with other early risers—stinging nettles and fiddlehead ferns. After a long Pacific Northwest (or north-*wet*) winter, I welcome rhubarb like a long-lost friend. I grew up eating rhubarb all through the spring and summer. Dad grew it in our small backyard garden, a callback to his family's southern Indiana farm.

Related to the celery family, rhubarb plants are perennial and will go dormant each winter, emerging slowly in March, sending out leaves first and then long stalks of red or green. Green stalks are not unripe, but just a different variety of rhubarb. Just south of Seattle in the Fife-Sumner valley is the rhubarb capital, where the most rhubarb in the United States (and possibly the world) is grown. (Check out the Rhubarb Compendium at RhubarbInfo.com.) In late winter and very early spring, before field rhubarb is available (and I can't wait any longer), I treat myself to the hothouse variety, which has its own flavor profile and brilliant pink interior. The best way to freeze rhubarb is in one-pound storage bags, cut into the piece size you'll use later.

Vanilla Roasted Rhubarb

Rhubarb and strawberries are my favorite fruits for roasting. Strawberries work just as well in this recipe. Roasting fruit enhances and deepens flavor, bringing out its caramel notes. Roasting can also bring back to life some less-than-prime or slightly unripe fruit. You can achieve the best flavor and most intense color by holding roasted fruit in the refrigerator overnight. The brilliant red and pink colors and deep flavor transform with time. The roasting juices also make an excellent flavoring for seltzer or your favorite vodka.

VARIATION

Vanilla Roasted Strawberries

You can use this same roasting technique for strawberries (and many other fruits, such as plums, peaches, cherries, apples, and pears). Hull the strawberries and halve the larger berries, keeping the smaller berries whole (so you have similar-size pieces), and roast according to the instructions. Since strawberries are sweeter than rhubarb, you may find you can use less sugar.

Makes 4 servings

6 tablespoons sugar (or less, if you prefer)

1 vanilla bean, split and scraped, or ½ teaspoon ground vanilla bean, or 1 teaspoon vanilla bean paste

Pinch kosher salt

4 cups (454 g) of 1-inch rhubarb pieces (from about 4 stalks)

1. Preheat the oven to 350 degrees F (175 degrees C).

2. Combine the sugar, vanilla bean pulp and pod, and salt in a large bowl, and use your hands to work the vanilla into the sugar. Add the rhubarb and toss to coat. Spread the rhubarb on a rimmed baking sheet.

3. Roast the rhubarb for 15 to 20 minutes, gently stirring or shaking the pan halfway through, until the fruit is tender to the touch but not falling apart. (Stir the fruit gently one time.) The roasting time will change depending on the size of your fruit pieces.

4. Cool the rhubarb to room temperature, and serve it with the extra juices. The fruit will hold refrigerated for 5 days.

Rhubarb-Ginger Sorbet

Sorbets are super simple—you don't need an ice-cream maker to make sorbet (or ice cream)—and they beautifully showcase all types of fruit. A basic stewing technique applies just enough cooking to soften the rhubarb and infuse the ginger, allowing the fruit to come through clean and bright. The rhubarb should be the star here, so don't get carried away with the ginger.

1. Make a sugar syrup by combining the sugar and ¾ cup (171 mL) water in a small saucepan. Cook the syrup over high heat until it boils. (The syrup can be made ahead and kept refrigerated for up to a week.)

2. Combine the rhubarb, fresh ginger, and the remaining ¼ cup (63 mL) water in a medium pot. Bring to a simmer, covered, on medium heat until the rhubarb is tender, about 15 minutes. Remove four of the eight pieces of ginger and discard. Combine the cooked rhubarb, remaining ginger, and the sugar syrup in a blender and puree until smooth. Pour the mixture into a metal bowl and chill until very cold—at least 2 hours. See How to Freeze Sorbet, Ice Cream, and Sherbet (page 8) for freezing instructions.

3. Serve in chilled bowls, garnished with the candied ginger. →

VARIATION

Fruit Sorbet

This simple sorbet formula allows you to replace the rhubarb with most any fruit. It's best to follow the weight of the fruit needed to get the best ratio of fruit to simple syrup, but any fruit cut into ½-inch pieces can be measured by volume. Strawberries, raspberries, plums, peaches, apricots, and tart apples work very well with this recipe.

Makes 6 servings

1 cup (200 g) sugar (see note, page 8)

1 cup (227 mL) water, divided

4 cups (454 g) of ½-inch rhubarb pieces (from about 4 stalks)

1-inch piece fresh ginger, peeled and cut into 8 slices

1 tablespoon candied ginger strips (optional)

How to Freeze Sorbet, Ice Cream, and Sherbet

Freeze the sorbet, ice cream, or sherbet in an ice-cream machine according to manufacturer's instructions. No ice-cream machine? No problem. Just briskly whisk the chilled mixture about ten times, and place it in the freezer. Every 20 to 30 minutes, remove the mixture and whisk ten more times. Then return to the freezer. Expect to whisk the mixture four to eight more times over the next 2 to 4 hours, depending on how fast your freezer chills it. (Each time you whisk the sorbet, ice cream, or sherbet you add air into it, which gives it the appropriate texture.) Continue this process until the consistency is like very soft soft-serve ice cream. Next, transfer the mixture to an airtight container and freeze it at least 4 hours or overnight.

Note: Since sugar is a liquefier in recipes, it provides not just sweetness but also texture and, in this case, pliability. Sugar is what makes sorbet and ice cream scoopable. So if you cut back on the sugar, your sorbet will be icier and, overall, harder to scoop. I always recommend cutting back sugar in increments—reducing by one-quarter or one-third first.

Straight Rhubarb Pie

This is the rhubarb pie I grew up eating. There are no strawberries in this pie. I can still see my mom leaning out the glass slider to the back porch, calling, "Larry, bring me some rhubarb for a pie!" Dad always complied because Mom is a legendary pie maker, and no one would want to keep her from making one. When Mom first tasted this pie, she looked at me skeptically, and I had to come clean on the addition of orange zest. You may say yea or nay to this flavor, so try it both ways. Mom did and liked it.

Makes 6 servings (or one 9-inch pie)

1 recipe Flaky Pie Dough (recipe follows)

7 cups (840 g) rhubarb pieces (from about 7 stalks, cut ¾-inch long)

¾ cup (150 g) sugar, plus more for finishing

⅓ cup (44 g) all-purpose flour

½ to 1 teaspoon orange zest

1 vanilla bean, split and scraped, or ½ teaspoon ground vanilla bean, or 1 teaspoon vanilla bean paste

¼ teaspoon kosher salt

1½ tablespoons unsalted butter

2 tablespoons whole milk, for finishing

Vanilla ice cream or whipped cream (optional)

1. Prepare the pie dough.

2. Combine the rhubarb, sugar, flour, orange zest, vanilla bean pulp, and salt in a large bowl. Toss until the vanilla and orange zest is distributed and all the fruit is coated with sugar and flour. Set aside.

3. Roll out one disk of the pie dough for the bottom crust to about ⅛ inch thick and line a 9-inch pie pan. (See How to Roll Pastry [Pie] Dough, page 13.) Trim the edge of the bottom crust with scissors or a knife to ¼ inch over the edge of the pie pan. Toss the rhubarb filling a few more times to pick up sugar sitting on the bottom of the bowl, and spoon the rhubarb filling into the pie pan, mounding the filling 1 to 2 inches above the edge of the pie pan. Break up the butter in nickel-size pieces and distribute over the top of the rhubarb. →

(If you have extra sugar and flour at the bottom of the bowl, just sprinkle it over the top of the fruit in the pie pan.)

4. Roll out the second dough disk for the top crust to about ⅛ inch thick. Place the crust on top of the pie, and trim the edge flush with the bottom crust. Pick up the overhanging edges, and turn them under all the way around the pie, pressing the dough onto the rim of the pie pan, or just inside the pie pan if your pan doesn't have a flat edge. Now crimp the edge with a fork or between your fingers to create a seal and a pretty edge.

5. Chill the finished pie in the freezer for 30 minutes before baking. This step will keep the dough from melting before it bakes and help achieve a baked crust on the bottom. The pie crust should be very cold and firm before going into the oven. While the pie is in the freezer, preheat the oven to 400 degrees F (200 degrees C).

6. Before baking, brush the top of the pie with milk and sprinkle sugar over the top. Bake the pie for 20 minutes or until the top crust is dry and just starting to color. Reduce the oven temperature to 350 degrees F (175 degrees C) and continue baking for 30 to 40 minutes more, or until the pie is brown and bubbling. The pie should be golden on top, and you should be able to hear or see the juices bubbling in the middle of the pie. (If you're unsure whether the fruit is cooked, insert a dinner knife into one of the slits. No resistance means the fruit is soft and cooked.)

7. Cool the pie to room temperature and serve with vanilla ice cream or whipped cream.

FLAKY PIE DOUGH

Makes two 9-inch pie crusts

2½ cups (325 g) all-purpose flour
1 teaspoon kosher salt
1 cup (2 sticks; 227 g) unsalted butter, cold, cut into ½-inch pieces
½ cup (113 mL) ice water, plus more as needed

- Combine the flour and salt in a large bowl. Add the butter pieces to the flour, and quickly work it in with your hands by pinching the butter and flour between your fingers, working the butter into the flour. Continue tossing and processing the dough and butter with your fingers until you still have some larger pieces of butter (dime-size) and the rest of the flour looks like cornmeal from the butter being worked into it.

- Add ½ cup of the ice water. Using a fork, toss the ingredients together, cutting through the crust mixture several times. Continue to add 1 tablespoon of ice water at a time until the dough starts to hold together. The dough should look shaggy and not come together easily in the bowl. Test the dough's moisture by gently squeezing it with your hand: if it holds together, it's ready; if not, add a little more water until it holds.

- Place the dough onto a work surface and quickly gather it together with your hands into a rough disk. Fold the dough in half, flatten it with your hand, and fold it a second time. The dough should be holding together at this point. (If not, sweep the dough back into the bowl, break it apart, and add another tablespoon of water. Don't fold it again, as you may overwork the dough.) Divide the dough into two equal parts. *Gently* form two disks about 1 inch thick. Don't work the dough or try to make it pretty. Small cracks and chunks of butter are normal. Wrap the disks with plastic wrap and chill in the refrigerator for 1 hour or overnight.

- The dough disks can be double wrapped and frozen for up to 4 weeks; thaw them overnight in the refrigerator before using.

VARIATION

Whole Wheat Pie Dough

You can replace the all-purpose flour with white pastry flour or whole wheat pastry flour. Note that whole wheat pastry flour will require less water. Start with ¼ cup of water and then add 1 tablespoon at a time until the dough holds together.

How to Roll Pastry (Pie) Dough

Remove the chilled disks from the refrigerator, and roll out each disk with a rolling pin on a surface dusted lightly with flour. (If your dough has been in the refrigerator overnight, it probably needs 5 to 10 minutes on the counter at room temperature to become pliable before rolling.) Dust the top of the dough with flour before rolling. I like to pound the dough with the rolling pin three or four times, then turn it one-quarter turn and pound it again five or six times to start to soften it. Then roll about six times from the middle up and middle down. Turn the dough a quarter turn (grabbing some of the flour on the counter for the bottom and dusting the top again if needed) and continue to roll six times up and down. (Remember to only add flour as needed.) Continue this pattern until the dough is about ⅛ inch thick. Use your hands to feel the dough all the way around to ensure a consistent thickness. (On a hot day, the dough may become too warm and soft while rolling it out. Return it to the fridge for about 15 minutes to chill, and you get to take a chill break too.)

Rhubarb Fool

Such a simple and yet versatile dessert, this fool depends on a gentle hand and a pretty, clear dish or glass. Originally, fools were made with a cooked custard folded into the prepared fruit. Over time, the recipe has evolved into this modern version, replacing the custard with whipped heavy cream. Tart fruits work best in this dessert because you need a deep, sharp flavor to shine through the richness of the heavy cream. Gooseberry fool is the most traditional, but rhubarb, strawberry, and raspberry are also common. For an even grander dessert, make the Vanilla Roasted Rhubarb (page 5) and add it to the layering in the final presentation.

VARIATION

Infused Rhubarb

For a change in flavor, add a strip of orange zest or a few slices of fresh ginger to the cooking rhubarb. Remove them after cooking, or continue to let them infuse while the rhubarb cools. Remove any flavor infusions before combining the compote with the whipped cream.

Makes 6 servings

3 cups (340 g) ½-inch rhubarb pieces (from about 3 large rhubarb stalks)

⅓ cup (67 g) sugar

⅓ cup (76 mL) water

½ vanilla bean, split and scraped, or ¼ teaspoon ground vanilla bean, or ½ teaspoon vanilla paste

1½ cups (340 mL) heavy cream

1. Combine the rhubarb, sugar, water, and vanilla bean pulp in a medium pot. Bring the fruit to a simmer over high heat and then reduce the heat to low (but still gently simmering). Cover the pot and cook the rhubarb until it is very tender and breaks apart, about 8 minutes. Remove the cover and increase the temperature to medium, simmering the rhubarb until it thickens and the liquid is reduced, about 5 minutes. Transfer to a large bowl, and chill the rhubarb until it is cold. (The rhubarb compote will thicken and the color will intensify while it is chilling.)

2. Whip 1 cup (227 mL) of the heavy cream to medium peaks with a mixer fitted with the whisk attachment. Reserve 1 cup (227 mL) of the rhubarb compote, and *gently* fold the whipped cream into the remaining chilled rhubarb.

(Don't overmix, or the fool will become runny.) Divide half the fool among six clear parfait glasses or dessert bowls. Top the fool in each dish or glass with a tablespoon of the reserved rhubarb compote.

3. Add the remaining fool to the dish or glass and then top again with the remaining rhubarb compote so that you have a layered effect of fool, compote, fool, compote. (If you're comfortable using a piping bag, it's an easier way to get the components into a narrow parfait glass.) Chill the fool for at least 1 hour before serving. Finish by whipping the additional ½ cup (113 mL) heavy cream and spooning or piping a dollop on top of each dessert. Serve this dessert cold.

STRAWBERRIES

For three weeks every June, after some long-awaited sunshine, the farmers' markets and fruit stands in western Washington are overflowing with local strawberries. Although acceptable strawberries are available all year round, I yearn for the local strawberries available in those fleeting weeks. These are not the strawberries that will hold in a fridge for four or five days. Instead, these berries must be used the same day or next day because after that, they will have sadly molded or turned to mush. The flavor and texture are like no other strawberries I can find the rest of the summer.

We picked strawberries as youngsters with my mom at farms near Carnation in the Snoqualmie Valley, at the base of the foothills of the Cascade Range. Fellow pickers know it's back-breaking work, leaning over to reach low plants, often on muddy ground in the rain, as we frequently have here in western Washington. But these strawberries are worth it. They make superior strawberry shortcake and jam, and are fabulous roasted, as I mentioned in the Vanilla Roasted Rhubarb variation in the previous chapter (page 5).

Strawberry Pavlova

The pavlova has a romantic history. Created by an Australian pastry chef for Russian ballerina Anna Pavlova while she was performing in Australia in the 1920s, the dessert was described by the chef as being as light as she was. A traditional pavlova is topped with fresh strawberries, kiwi, and passion fruit. Although I'm not generally a fan of eating hard-baked meringues, I thoroughly enjoy this dessert with its combination of flavors and textures—tangy fruit, rich whipped cream, and crisp, chewy meringue. I have also paired this pavlova with roasted rhubarb—a perfect combination.

VARIATION

Family-Style Pavlova

For a family-style dessert to bring to the table, make one large pavlova, about 9 to 10 inches round. You will need to bake the large version for 60 to 70 minutes. Don't expect to cut this grand creation into nice slices. Just use a large spoon to portion it at the table.

Makes 6 to 8 servings

1 recipe Vanilla Roasted Strawberries (page 5)

For the meringues:
1½ cups (300 g) sugar

1½ tablespoons cornstarch

6 egg whites at room temperature

½ teaspoon cream of tartar

¼ teaspoon kosher salt

2 teaspoons vanilla extract

For the whipped cream:
1½ cups (340 mL) heavy cream

2 tablespoons sugar

1. Prepare the roasted strawberries.

2. *To make the meringues,* preheat the oven to 275 degrees F (135 degrees C). Line a baking sheet with parchment (add a little nonstick spray or butter underneath to keep the parchment from slipping).

3. Whisk together the sugar and cornstarch in a small bowl.

4. Combine the egg whites and cream of tartar in the bowl of a stand mixer with the whisk attachment. Whip on medium-high speed until the egg whites are just beginning to foam. With the mixer running, add the sugar-cornstarch mixture in four parts, waiting about 5 seconds between each addition, until all of it is has been added. (Adding the sugar too fast will deflate the egg whites, but adding the sugar too slow won't →

allow the sugar to dissolve correctly.) Turn the mixer to high speed, and continue whipping the meringue to stiff peaks. The meringue should be shiny and create a firm pattern in the bowl. Stop the mixer and test the meringue by dipping your finger into it. The peak should stand straight up on your finger. Add the salt and mix at medium-high speed for 3 minutes. This stabilizes the meringue and helps the sugar dissolve. Add the vanilla and mix at medium speed.

5. Portion the meringue into six mounds about 1 inch apart on the baking sheet. Using a tablespoon, spread each meringue to about 4 to 5 inches in diameter, making a divot ¾ inch deep by 2 inches wide in the middle of each meringue (to hold the cream and fruit later). You can make any size meringues, but remember that smaller meringues will take less time to bake.

6. Bake the meringues for 40 minutes, or until they are dry and crisp on the exterior. The meringues will feel dry after about 30 minutes, but continue baking them another 10 minutes to ensure the interior is baked. If the meringues start to color, lower the temperature to 250 degrees F (120 degrees C). Keep in mind that in wet or humid weather, meringues can be finicky. You may need to bake the meringues longer to get to a crisp stage. Pavlova meringue interiors should be soft, like marshmallow. Cool the meringues at room temperature. If you are going to store them, move them to an airtight container and stack the meringues between parchment as soon as they are cool. They will hold for about 2 days. (Moist air will cause the meringues to soften.)

7. When you are ready to serve, whip the cream with the sugar to medium-stiff peaks. Put a dab of whipped cream on each plate, and then place the meringues on the plates. (The whipped cream is a plating trick to keep the meringue from sliding around as you take it to the table.) Fill the indentations of each meringue generously with whipped cream. Divide the roasted strawberries among the pavlovas, drizzling the juice over the top, and serve immediately. You can also build the pavlovas ahead and refrigerate them for up to 30 minutes before serving.

Strawberry-Lime Layer Cake

One of my most requested desserts, this cake is simple to make and a joy to eat. All the components can be made up to 3 days ahead, but it's also not too overwhelming to make them all in the morning—build the cake and let it chill at least one hour. A perfect genoise, stepped up with a rich mascarpone filling, roasted strawberries, and a hint of lime (either candied or fresh). The flavors meld nicely to create a lovely spring dessert, worthy of showcasing at the table before you cut into it.

Makes 8 servings

1 recipe Quick Candied Lime Zest (recipe follows) or fresh zest from 2 limes

For the genoise cake:
3 eggs

3 egg yolks

½ cup plus 5 tablespoons (163 g) granulated sugar

2 tablespoons vanilla extract

3 egg whites

½ cup (65 g) all-purpose flour, sifted

1 recipe Vanilla Roasted Strawberries (page 5)

For the filling:
2 cups (454 g) mascarpone

½ cup (114 g) sour cream

1 cup (114 g) confectioners' sugar, sifted

Zest from 2 limes

6 tablespoons freshly squeezed lime juice (from about 5 limes)

⅓ cup (38 g) confectioner's sugar, for finishing

1. Prepare the candied lime zest.

2. *To make the cake,* preheat the oven to 400 degrees F (200 degrees C). Lightly grease the bottom of a 12-by-16-inch baking pan, and lay parchment on the bottom. (Be sure to trim the parchment so it lies flat.)

3. Combine the eggs, egg yolks, 5 tablespoons of the sugar, and vanilla in the bowl of a stand mixer fitted with the whisk attachment. Whip on high speed until the mixture is pale yellow, thick, and forms a ribbon, 5 to 8 minutes. Move the mixture to a large bowl, and clean the mixer bowl and whisk for the next step.

4. Whip the egg whites in the bowl of a stand mixer with the whisk attachment on high speed, until they are white and frothy. With the mixer running on high speed, →

add ½ cup (100 g) sugar in three parts. Continue whipping on high speed until the egg whites hold a medium peak. If you stop the mixer and dip your finger into the whipped whites, the peak should curl sideways (see note, page 25).

5. Using a rubber spatula, fold one-third of the meringue into the egg yolk batter using a gentle folding motion. Once the first third of the whites is mostly incorporated, fold in the remaining whites until completely combined. Be careful to not overmix the batter or you will deflate the whites.

6. Fold in the flour in three parts. Be sure to cut through the center of the batter a few times to ensure the flour is well combined. Again, do not overmix.

7. Pour the batter into the prepared pan. Using an offset metal spatula and a light touch, spread the batter to the edges, making sure it is level. Gently tap the pan on the counter once to release any air bubbles. Bake the cake for 12 to 15 minutes, until the cake is light golden and springs back when touched. Remove the cake from the oven and cool in the pan. At this point, the cake can be triple wrapped with plastic wrap and frozen for up to 1 month.

8. While the cake is baking, prepare the strawberries for roasting. After removing the cake from the oven, reduce the temperature to 350 degrees F (175 degrees C) and roast the strawberries.

9. Once the cake is cool, loosen the edges from its pan with the edge of a plastic bowl scraper or a knife. Lift the cake (with attached parchment) out of the pan and flip it over, top down, onto a clean piece of parchment. (Genoise cakes are surprisingly strong and flexible due to the large amount of eggs.) Remove the parchment from the back of the cake by gently peeling it back, starting at one corner and moving to the opposite corner; hold the cake with one hand to keep it from moving with the parchment. Using a pastry wheel or a knife, trim the edges of the cake about ½ inch all the way around. Then divide the cake into three equal pieces, each 5 inches wide.

10. *To make the filling,* combine the mascarpone and sour cream in the bowl of a stand mixer fitted with the paddle attachment on low speed (or by hand),

scraping once. You will mix the filling three times total, after three different additions. Turn the speed up to medium and mix until the filling thickens and holds a peak, 30 to 45 seconds. Add the sifted confectioners' sugar and mix on low speed until combined, scraping once. Turn the speed up to medium and mix until the mixture thickens, another 30 to 45 seconds. (Be careful not to overmix, or the filling will become loose. If this happens, refrigerate it for 30 minutes, then gently mix it until smooth.) Add the lime zest and juice and mix on low speed until combined. Turn the mixer up to medium speed again and mix until the filling stiffens a third time, 30 to 45 seconds. The consistency should be like thick, heavy whipped cream that holds a peak.

11. *To assemble the cake,* drain the roasted strawberries (reserving the juice for serving). Lay one cake layer on a rectangular plate. Spread half the mascarpone filling on the cake. (If the filling has been chilled, it may be stiff. Give the filling a few good stirs, and let it sit at room temperature for 5 to 15 minutes. Make sure it is pliable enough to spread without tearing the cake.) Distribute half the drained strawberries evenly on top of the mascarpone filling, pressing the strawberries into the filling. Place the second cake layer on top of the strawberries (using your hands to gently level the cake), and repeat the steps with the second half of the mascarpone filling and the strawberries. Place the third cake layer on top, pressing it gently to secure it. Carefully clean up the sides of the cake with a knife, leaving them open to expose the layers. Refrigerate the cake for 2 hours. Sift the confectioners' sugar over the cake, and sprinkle the top with the candied lime zest or fresh zest from 2 limes.

12. Use a serrated knife to cut the cake into servings, and drizzle some of the leftover strawberry syrup alongside each piece of cake.

Note: There are three general stages of whipped egg whites—soft, medium, and stiff. Soft peaks curl over and touch the meringue below. Medium peaks curl sideways, but hold there. Stiff peaks point straight up.

QUICK CANDIED LIME ZEST

Makes about 2 tablespoons candied zest

2 limes
2 tablespoons sugar

- Using a vegetable peeler, strip the zest from the limes starting at the stem end and ending with the bottom of the lime. Cut the zest strips into the thinnest-possible long strips. Toss the cut zest strips and sugar on a parchment-lined baking sheet, rubbing them together with your hands.

- Spread the zest and sugar on the baking sheet and let dry for 2 hours or overnight. The zest is ready when it is crunchy and stiff. Store the dried zest in an airtight container at room temperature. It will last for weeks if kept in the container. This is a simple way to create decorative zest that looks beautiful and tastes good too.

Strawberry-Coconut Sorbet

Strawberries may be the most well-liked of all fruit. Their ability to be paired with many other flavors makes them even more loved than just being lovable on their own. This sorbet came to me some years back during the height of our short strawberry season when fresh strawberries didn't need any cooking to enhance them. The relatively large ratio of honey to granulated sugar means the honey is a substantial player here. Try a local honey to change up the flavor, but be choosy: the stronger the honey, the more it will influence the flavor of the sorbet.

Makes 6 servings

1 cup (240 mL) coconut milk

¼ cup (55 mL) fireweed or other mild honey

2 tablespoons sugar

¼ teaspoon kosher salt

½ vanilla bean, split and scraped, or ¼ teaspoon ground vanilla bean, or ½ teaspoon vanilla bean paste

4 cups (600 g) hulled and quartered strawberries (about 3 pints)

¼ teaspoon coconut extract

3 tablespoons large-shaved dried coconut, toasted, for garnish (see note, page 29)

1. Combine the coconut milk, honey, sugar, salt, and vanilla bean pulp and pod in a medium pot. Bring to a boil over high heat, stirring occasionally. Remove the pot from the heat, and let the coconut mixture steep for 20 minutes.

2. Put the strawberries into a blender. Remove the vanilla bean pod from the coconut milk mixture, and pour about half the coconut mixture into the blender with the strawberries. Cover the blender container tightly with the lid (remove the small plastic cover piece from the center of the lid, and cover that area loosely with a thick towel, holding it gently with your hand to let out steam). Process on low speed for about 10 seconds. Turn the blender to high speed and process until smooth, about 30 seconds.

3. Pour the sorbet mixture through a fine-mesh sieve to remove the strawberry seeds. (Use a rubber spatula to actively stir and push the sorbet mixture through the sieve.) Add the remaining coconut mixture and the coconut extract to the sorbet mixture.

4. Chill the sorbet mixture until it is cold, about 2 hours or overnight. See How to Freeze Sorbet, Ice Cream, and Sherbet for freezing instructions (page 8).

5. Divide among six cold dishes and serve topped with toasted coconut.

Note: To toast the coconut, preheat the oven to 350 degrees F (175 degrees C). Place the coconut on a parchment-lined baking sheet. Bake the coconut for 5 to 8 minutes, or until golden brown, stirring as needed to toast the coconut evenly.

< Strawberry-Coconut Sorbet ^ Roasted Strawberry Ice Cream

Roasted Strawberry Ice Cream

There are those who like their strawberry ice cream pink and those who want it chunky with strawberries. This recipe satisfies both. Roasting fruit is a gentle cooking method that draws out flavor and provides some caramelization of sugars. Though I've tried this recipe using crushed fresh strawberries and also stewed strawberries, neither preparation was as unique or deeply satisfying as this roasted-strawberry version. In the words of one of my students, "I've never tasted any ice cream quite like it." I don't think this ice cream needs any accompaniments. However, for those readers who think more is more, make a full batch of the Vanilla Roasted Strawberries (page 5) rather than just the half recipe that this recipe calls for, and use the extra for topping your already-very-strawberry ice cream.

Makes about 1½ quarts (about 1½ L)

3½ cups (800 mL) half-and-half

1¼ cups (250 g) sugar

⅓ cup (75 mL) honey or corn syrup

2 vanilla beans, split and scraped, or 1 teaspoon ground vanilla bean

10 egg yolks (about 210 mL)

½ recipe Vanilla Roasted Strawberries (page 5)

1. Combine the half-and-half, sugar, honey, and vanilla bean pulp and pod in a medium pot. Heat the cream mixture over high heat until it is hot. (If it's too hot to touch, it's hot enough.) Do not allow the cream to boil. Cream that has boiled will overcook this custard-style ice cream. If the cream boils, pour it into a metal bowl and let it cool for about 15 minutes. Then return it to the pot and heat to the desired temperature again.

2. While the cream mixture is heating, whisk the egg yolks in a medium bowl. Set up a fine-mesh sieve or a sieve with four layers of cheesecloth over a medium stainless bowl for straining the mixture.

3. When the cream mixture is hot, turn the heat down to medium-high and temper the cream mixture into the yolks. This is done by whisking the yolks while slowly adding about half the hot cream into the yolks. Next, transfer the egg yolk and hot cream mixture to the pot while whisking in the pot, combining all the egg yolks and cream. Cook over medium-high heat, stirring constantly with a spatula. Use a stirring pattern of continuously moving the blade of the spatula around the corners of the pot and in an S pattern across the bottom so that no spot is missed. Continue cooking until the mixture visibly thickens and coats the back of the spatula—also called *nappé* in French. Test this by drawing your finger across the spatula, creating a definite line that does not run together. (If your cream was on the hotter side when you added the eggs, you will reach nappé more quickly—within 2 to 3 minutes.) Immediately remove the mixture from the heat, and strain it through the fine-mesh sieve or cheesecloth. (Overcooking the custard will cause the eggs to cook, and the mixture will look curdled.)

4. Cool the mixture in the refrigerator until cold, 2 hours or overnight. Meanwhile, prepare the roasted strawberries.

5. Puree the roasted strawberries in a blender, leaving some texture and small chunks. (If you like bigger chunks of fruit, chop some of the strawberries by hand and puree the rest.) Stir all the strawberries into the chilled ice cream. (Refrigerating the finished ice-cream base overnight before freezing it will improve the flavor. The ice-cream base can be held in the refrigerator for 3 days, or you could freeze the base as is for later use.) See How to Freeze Sorbet, Ice Cream, and Sherbet for freezing instructions (page 8).

SUMMER

CHERRIES 37

Cherry Hand Pies 39

Smoky Sweet Cherries in Port with Bittersweet Chocolate 41

Sour Cherry Compote 45

White Chocolate Mousse with Poached Sweet Cherries 46

RASPBERRIES 51

Raspberry and Lemon Balm Shortcake 52

Port Sabayon with Fresh Raspberries 55

Raspberry Custard Tart 56

Dark Chocolate Cake with Raspberry-Orange Compote 62

BLUEBERRIES 65

Classic Lattice Blueberry Pie 67

Blueberry-Cinnamon Crepes 70

Blueberry and Lemon Curd Tart 76

Poached Blueberries with Vanilla Bean and Anise 79

BLACKBERRIES AND MARIONBERRIES 81

Marionberry Crostata with Whole Grain Crust 82

Pain Perdu with Assorted Berries and Grand Marnier 86

Bittersweet Chocolate Tart with Blackberries and Basil 89

Vanilla Bean Cake with Glazed Blackberries and Stone Fruits 91

APRICOTS 97

Grilled Apricots with Brown Butter and Maple-Tamari Glaze 98

Apricot and Walnut Streusel Bars 102

Poached Apricots in Sauternes 105

PEACHES AND NECTARINES 109

Peaches and Nectarines Sautéed with Thyme 110

Peach Mousse 114

Roasted Peach Bread Pudding 116

CHERRIES

Although Washington is known as the apple state, each July it feels like the cherry state. (It is not. Michigan easily holds that honor.) Washington produces about twenty-eight million pounds of Bing, Rainier, and other types of sweet cherries, along with much smaller crops of tart cherries. Cherry growers wait nervously just before the cherries are harvested, hoping it won't rain and cause the cherries to split. Crops have been ruined over our rain. I listen for the cherry crop predictions like a wheat farmer listens for the wheat prices. Although the Rainier cherry is coveted for its beauty and fresh flavor, the Bing and other dark red cherries (such as Chelan, Tieton, Sweetheart, and Lapins) remain my favorites. I savor their deep red color and even deeper flavor for both fresh eating and cooking. Of all the fruit my mother canned throughout the summer, the Bing cherries were saved for a special occasion.

Sour cherries, like Montmorency (a type of morello cherry), are harder to find but worth the search. Their tartness packs a punch in a pie or sauce, where some sugar mellows their astringency. Cherries, like berries, are best kept refrigerated in a single layer, or pitted and frozen as soon as possible. A dip in a lemon juice and water mixture will keep pitted cherries from oxidizing in the freezer.

Cherry Hand Pies

Hand pies are simple to make and even easier to eat. With so many pie shops popping up, takeaway hand pies have become popular. They also showcase your pie crust because the ratio of crust to filling is much higher. Bite into one of these babies, and you'll experience the super-flaky, all-butter crust and a burst of either tart or sweet cherry, along with vanilla and almond. You can make them in any size (kids love the tiny ones), but these are a good size for a picnic or backyard barbecue. Don't stress about the pies leaking a little—they always do. Just be sure to grease the parchment, and go forth and make pie.

Makes 12 hand pies

1 recipe Flaky Pie Dough (page 12), divided into three equal disks

For the filling:

3½ cups (530 g) pitted and halved sweet or sour cherries

¼ cup (50 g) sugar for sweet cherries, or ½ cup (100 g) sugar for sour cherries, plus more for finishing

3 tablespoons cornstarch

2 teaspoons freshly squeezed lemon juice or apple cider vinegar

½ vanilla bean, split and scraped, or ¼ teaspoon ground vanilla bean, or ½ teaspoon vanilla bean paste

½ teaspoon almond extract (optional)

2 pinches kosher salt

For the egg wash:

1 egg

1 tablespoon water

Pinch kosher salt

2 tablespoons whole milk, for finishing

1. Prepare the pie dough.

2. *To make the filling*, combine the cherries, sugar, cornstarch, lemon juice or vinegar, vanilla bean pulp, almond extract (if using), and salt in a bowl and toss gently to combine.

3. *To make the egg wash*, combine the egg, water, and salt. Whisk briskly until the egg is thoroughly broken up.

4. Roll out one disk of dough to about 9 inches square. (See How to Roll Pastry [Pie] Dough, page 13.) Use a pastry wheel or knife to square up the dough by conservatively trimming the edges to make a square. Now cut the trimmed dough through the center into four equal squares. →

5. Turn over each square of dough so that the more floured side is facing up. Lightly brush egg wash along the half-inch outside edge of each dough piece. Place the filling on one side of the dough. Fold the dough over the filling, corner to corner, attaching the center point to make a triangle. Then seal along one side, using one hand to gently keep the fruit inside the pie. Repeat on the other side. I'm the queen of overfilling my hand pies. They need to be full, but not so much that you can't close them. Inevitably, I end up removing fruit as I start to seal the edges so that I can get them closed. Repeat with all four squares.

6. Once the pies are sealed, crimp the edges with a fork. With juicy fruit, you will notice some leaking. Do your best to seal the pies, chill them for 10 minutes, and then seal them again with the fork. Cut a few small vents in each pie and place them, spaced evenly, on two parchment-lined baking sheets.

7. Roll out the second and third dough disks, repeating with the remaining filling. Freeze the pies for 30 minutes. (After freezing, you can store them in a bag or container in the freezer for 2 to 4 weeks to bake anytime directly from the freezer.) While the pies are in the freezer, preheat the oven to 400 degrees F (200 degrees C).

8. When you are ready to bake, brush the top of the pies with milk and sprinkle with sugar. Bake the pies for 10 to 15 minutes (or until they are dry and starting to brown). Reduce the temperature to 350 degrees F (175 degrees C) and continue baking the pies until they are golden brown and the juices are bubbling, 10 to 15 minutes longer.

9. Cool the pies to room temperature or barely warm before eating out of hand. Store the pies in an airtight container at room temperature for 2 days.

Smoky Sweet Cherries in Port with Bittersweet Chocolate

It's not often that the grill comes to mind for cooking fruit. Large, fleshy red cherries, such as Bing cherries, hold up well on the grill and take on a little smoke with some intense heat, and port deepens the flavor. This isn't a fancy dessert. I usually prepare it when I'm planning to barbecue, because the foil pans are easily moved to a picnic table, where guests can just pick up a few cherries by their stems.

Makes 6 servings

1 cup (227 mL) ruby port (tawny port will work here too)

2 tablespoons sugar

3 cups (454 g) whole sweet red cherries, stems on (not pitted)

¼ cup (57 g) large-grated dark chocolate (such as 70% cacao)

1. Combine the port and sugar in a small saucepan. Bring mixture to a boil over high heat, and then reduce the heat so that the port simmers until it is reduced to a syrup (by about two-thirds), about 20 minutes. (Once the port starts to reduce, stay close to monitor it. It's easy to burn the port syrup as it reduces. Test the consistency on a cold plate to make sure the port is the consistency of maple syrup.) The syrup can be made up to 1 week ahead and stored in the refrigerator. Bring it to room temperature before using.

2. Preheat the grill. If you are using a gas grill, set the heat on medium flame. However, all grills work differently, and charcoal grills vary widely, so use your intuition →

and adjust your grill as needed. You will want the cherries to cook and soften without burning and shriveling, so you may need to lower your heat substantially. (See note for oven options.)

3. Using two doubled pieces of heavy-duty foil, create two makeshift shallow pans, about 6 by 8 inches. Lightly grease the pans. Divide the cherries between the pans. Drizzle the port syrup evenly over the cherries. Place the foil pans of cherries on the grill and close the lid. When the syrup starts to bubble, turn the heat down to low and continue to cook the cherries, stirring occasionally—about 10 minutes, but this will vary depending on the type of grill you use. The cherries will soften with some dark spots.

4. Move the foil pans to a heat-resistant surface. Immediately sprinkle the grated chocolate over the top of the warm cherries and serve. You can leave the cherries on the pans and serve them casually, or move them to a plate or wide bowl.

5. These cherries are best served and eaten promptly while they are warm, but that never stops me from snacking on the leftovers a few hours later.

Note: If you don't have a grill, you can make these smoky cherries in your oven by using the broil setting. Instead of using the foil pans, line a rimmed baking sheet with foil and place all the cherries and port syrup on the baking sheet. Preheat the broiler for 5 minutes on high. Place the pan of cherries in the oven on the second-to-highest rack. Broil for about 4 minutes or until the port syrup boils. Open the oven and quickly shake or stir the cherries. Broil for another 4 to 5 minutes and then move the cherries down to a lower rack and continue cooking until they are soft (about 10 minutes), as described with the barbecue version above. Finish with the grated chocolate.

Sour Cherry Compote

This deeply flavored compote with gorgeous red color is a little sweet and a little sour, with three different sweeteners to tame the acidic nature of tart cherries. The first time I served this over ice cream alongside a competing Bing cherry compote, the sour compote won hands down. Keep it in the refrigerator to top a simple cake like the Almond Cake (page 173), to garnish pancakes, or to add to your favorite mixed drink.

Makes about 2½ cups (568 mL)

½ cup (100 g) granulated sugar

2 tablespoons demerara sugar or brown sugar

2 tablespoons honey

2 tablespoons Madeira or other fortified red wine, such as port

2 tablespoons water

½ vanilla bean, split and scraped, or ¼ teaspoon ground vanilla bean, or ½ teaspoon vanilla bean paste

Pinch kosher salt

3 cups (about 630 g) stemmed and pitted sour cherries, such as Montmorency (fresh or frozen)

1. Combine all the ingredients except the cherries in a medium pot and bring to a simmer over high heat, stirring a few times. Add the cherries and bring the compote back to a simmer, reducing the heat to medium-low until the cherries have softened some but are not falling apart, 15 to 20 minutes.
2. Cool and refrigerate the compote for up to 2 weeks.

Note: Madeira is a fortified wine produced in Portugal. Although it is a common cooking wine and should be available at all grocery stores, you could replace it with port or any other red wine.

White Chocolate Mousse with Poached Sweet Cherries

VARIATION

Dark Chocolate Mousse

Make it dark. Use this same recipe to make a dark chocolate mousse by replacing the white chocolate with 4 ounces (113 g) semisweet or bittersweet (60% to 75%) chocolate.

For those of you who don't like white chocolate, you probably haven't been eating *good* white chocolate. The key is the high percentage of cocoa butter (which is its only cacao ingredient). I'm a fan of Felchlin Edelweiss 36% white chocolate (available from some grocers and online). White chocolate can be quite sweet, so the pairing with the poached cherries cuts that sweetness and mellows the cherries at the same time. This mousse would also be a great vehicle for many other fruit preparations in this book.

For the poached cherries:
2 cups (454 mL) water

1½ cups (300 g) sugar

1½ cups (340 mL) brandy

1 vanilla bean, split and scraped

2 strips orange zest

¼ teaspoon kosher salt

1 star anise

3 cups (about 454 g) sweet red cherries, stemmed and pitted

For the mousse:
8 ounces (227 g) white chocolate, chopped

1 cup (227 mL) heavy cream

5 tablespoons sugar

1 egg

2 egg yolks

1. *To make the poached cherries*, combine the water, sugar, brandy, vanilla bean pulp and pod, zest, salt, and anise in a medium pot and bring to a simmer over high heat. Once the poaching liquid simmers, reduce the heat to low and continue to let it steep for 30 minutes to 2 hours. (The longer you let it steep, the stronger the flavors will be.) When you are ready to poach the cherries, gently drop them into the liquid and turn up the heat to medium. Once the poaching liquid returns to a simmer, remove the pot from the heat and pour the liquid and cherries into a large metal bowl or pan. The cherries

will continue to poach as they cool. Leave the cherries and liquid at room temperature until the liquid cools. Then move the bowl to the refrigerator and store overnight, if you have time. The overnight hold in the refrigerator will enhance the flavor and deepen the poached texture of the cherries. (I recommend removing the star anise after 1 hour in the refrigerator as the flavor can override the other infusions.) The cherries can be held in their liquid refrigerated in an airtight container or jar for 2 weeks.

2. *To make the mousse*, prepare a hot-water bath with 2 inches of water in a medium pot brought to a boil. Remove the water bath from the heat, and melt the white chocolate in a metal bowl set in the pan above the water. (The bowl should not touch the water.) Stir to melt the chocolate. Once the chocolate is melted, remove the bowl from the pot of water and keep the chocolate warm near the stove. Return the pot of hot water to the stove over low-heat for a future step.

3. Whip the heavy cream to medium-stiff peaks and refrigerate.

4. Combine the sugar, egg, and egg yolks in the bowl of a stand mixer. Rest the bowl on the water bath (the water should be lightly simmering, but not boiling), and whisk the eggs and sugar by hand until the mixture reaches 140 degrees F (60 degrees C). Move the bowl to the stand mixer fitted with the whisk, and whip the egg mixture until the bowl feels barely warm. (Cooling the egg foam will help stabilize the foam, keeping it from deflating.) Remove the egg foam from the mixer. Pour the melted chocolate into the egg mixture while gently folding with a spatula. (If a little chocolate hardens on the side of the bowl, just leave it there.) Fold just until the chocolate and egg mixture is combined. Temper the cream into the mousse by first folding in one-third of the cream. Then fold in the rest of the cream until the mousse is smooth and uniform. (Don't overmix or the mousse will deflate and become runny.) →

5. Divide the mousse among six 8-ounce glasses or dessert dishes. Chill them for at least 1 hour. If you make the mousse 1 day ahead, portion the mousse and stretch a small piece of plastic wrap over each glass and refrigerate.

6. Before serving, top the mousse with five or six cherries and a little drizzle of the cold poaching liquid. Serve immediately.

RASPBERRIES

Great raspberries are delicate. At their peak of ripeness, they barely make it from the vine to the kitchen without getting squashed. However, picking can be fast and easy, which means it's worth just camping out in the raspberry patch for a vine-to-mouth snack. But raspberries have a longer season than strawberries, and luckily, more growers are planting everbearing varieties that can take raspberry season from June all the way into late summer.

Raspberries may last a day or two in the fridge. My practice is to unbox them from the flat as soon as I bring them home. I transfer them to parchment-lined baking sheets in a single layer (searching carefully for any that might already be too soft or molding) and gently lay plastic wrap over the top. (Don't wrap them in plastic or they will build moisture in the fridge and spoil quickly.) When you're ready to use them, wash them very gently (a small handful at a time run under cool water) and lay them out on paper towels to dry.

Raspberries, unlike their delicate persona, pack a powerful flavor punch. This is the only chapter in the book where I've used the fruit almost exclusively fresh, instead of cooked. But they have been paired with several different ingredients to showcase their unmatchable flavor.

Raspberry and Lemon Balm Shortcake

Classics such as berry shortcake are hard to change up without disturbing traditional senses. Here, lemon balm—a seemingly mild herb that contributes immensely to flavor—is incorporated into the shortcakes for an herbal accent to the intense raspberries. Try infusing lemon balm into ice cream, fruit compotes and sautés, and even savory dishes.

Makes 8 servings

1 recipe Lemon Balm Scones (page 186; see variation)

½ cup (100 g) sugar

¼ cup (113 mL) mild honey

2 tablespoons maple syrup

2 tablespoons balsamic vinegar

8 cups (about 1 kg) fresh raspberries, gently washed and laid out to dry

2 cups (454 mL) heavy cream (for topping)

1. Combine the sugar, honey, maple syrup, and vinegar in a small saucepan. Heat on medium-low until the mixture is just warmed. Pour it over the raspberries and toss gently. Let the mixture sit at room temperature for 2 hours, tossing again two or three more times. (You can also make the raspberry mixture ahead and store it in the refrigerator up to 2 days. Bring the raspberry mixture to room temperature before serving.)

2. Prepare the shortcake biscuits.

3. When you are ready to serve, whip the heavy cream to medium peaks. Split each shortcake in half. Place the bottom piece in each serving bowl, and spoon about ½ cup raspberries over the shortcake. Top with the second shortcake piece, and spoon another ½ cup raspberries on top. Then spoon on a generous amount of whipped cream. Serve immediately.

Port Sabayon with Fresh Raspberries

A French classic, sabayon (or *zabaglione* in Italian) is so simple, I wonder why more cooks don't make it. I love the ease of spooning it over fresh fruit. The port makes it sweeter and slightly heavier than a traditional sabayon made with Marsala or white wine, and the lovely pink color is an attractive complement to the raspberries. For the fluffiest sabayon, make this up just before dinner and then let it continue to mix on low while you eat. The continual movement of the mixer preserves the volume of the sabayon.

Makes 6 servings

¾ cup (150 g) sugar

½ cup (113 mL) ruby port

6 egg yolks

Pinch kosher salt

6 cups (750 g) fresh raspberries, gently washed and laid out to dry for 30 minutes

1. Heat 2 inches of water in a medium pot over high heat to make a hot-water bath. Reduce the heat to keep the water at a low simmer. Combine the sugar, port, egg yolks, and salt in a stand mixer bowl, and set it over the water bath. (The bowl should not touch the water.) Whisk the mixture constantly until it reaches 165 degrees F (74 degrees C). Move the bowl to the mixer fitted with the whisk attachment, and whip the mixture on medium-high speed until the sabayon has cooled to room temperature. If there is no warmth on the bottom of the mixer bowl, it is adequately cooled.

2. Divide the raspberries among six bowls. Spoon the sabayon over the raspberries. If you have a torch at home, gratinée the sabayon by lightly torching it so that it caramelizes in spots. Serve immediately.

Raspberry Custard Tart

This is my favorite dessert, hands down. If I ever see one on a menu, I'm all in. One of the most traditional recipes in this book, this fruit tart is classic French pastry—Sweet Tart Dough (Pâte Sucrée)—baked well, filled with rich pastry cream, and topped with fresh raspberries and an apricot glaze. Be sure the tart shells aren't too thick and that they are baked dark golden so they hold up the filling.

Makes 6 servings (six 4-inch individual tarts or one 9-inch tart)

1 recipe Sweet Tart Dough (Pâte Sucrée) (recipe follows)

For the pastry cream:

¼ cup (50 g) sugar

2 tablespoons cornstarch

1 cup (227 mL) whole milk, divided

1 egg

2 egg yolks

⅓ cup (75 mL) heavy cream

½ vanilla bean, split and scraped, or ¼ teaspoon ground vanilla bean, or ½ teaspoon vanilla bean paste

For the apricot glaze:

¼ cup (113 g) apricot jam (do not use low-sugar or fruit-only jam)

¼ cup (57 mL) water

4 cups (500 g) fresh raspberries, gently washed and laid out to dry for 30 minutes

1. Prepare the tart dough, then bake the tart shells, up to 2 days ahead. Hold the baked shells in an airtight container.
2. *To make the pastry cream,* whisk together the sugar and cornstarch in a medium bowl. Add ½ cup (113 mL) of the milk and whisk until well combined, creating a slurry (or thickening agent).
3. Add the egg and egg yolks to the slurry and whisk until combined. Set a clean spatula and bowl next to the stove for your finished pastry cream.
4. Combine the remaining ½ cup milk (113 mL), heavy cream, and vanilla bean pulp and pod in a medium pot. Cook over high heat until the milk is steaming or boils. Temper the eggs by slowly pouring the hot milk into the egg mixture while whisking in the bowl. Transfer all the egg mixture back to the pot (off the heat), whisking in the pot until well combined.

5. Return the pastry cream to medium-high heat, then whisk constantly and vigorously until the mixture thickens and boils for 1 minute. (You will need to stop whisking for a few seconds to make sure it's boiling before you begin timing it.) Be sure to move the whisk around the corners of the pot and across the bottom regularly to avoiding scorching.

6. Immediately pour the finished pastry cream into the clean bowl and scrape with the clean spatula. (Always check the bottom of the pot before scraping to make sure you don't have any scorched spots. If you do, just avoid scraping there.) Press plastic wrap or parchment on the top of the pastry cream to prevent it from forming a skin. Refrigerate the pastry cream until it is cold. For extra flavor, I store the vanilla bean in the pastry cream until I've used the last of it. The pastry cream can be held in the refrigerator for up to 5 days.

7. *To make the apricot glaze,* combine the apricot jam and the water in a small pot. Heat on medium-high until the mixture simmers. Whisk well to combine and break up the jam. Strain the warm jam through a fine-mesh sieve and cool to room temperature before using.

8. *To assemble the tarts,* stir the pastry cream well, loosening it with a whisk or carefully with the stand mixer, using the paddle attachment. Divide the pastry cream between the tart shells, leaving ⅛ inch below the tops of each tart shell. Pile the raspberries on each tart, completely covering the pastry cream and creating a small dome of raspberries. With a pastry brush, gently glaze the raspberries so that they are completely covered with apricot glaze.

9. Chill the tarts up to 2 hours before serving.

SWEET TART DOUGH (PÂTE SUCRÉE)

Makes two 9-inch tart shells or six 4-inch tart shells

1 cup (2 sticks; 227 g) unsalted butter, room temperature
½ cup (100 g) sugar
1 egg, room temperature
1 teaspoon vanilla extract
2½ cups (325 g) all-purpose flour or pastry flour
1 teaspoon kosher salt

- Combine the butter and sugar in the bowl of a stand mixer fitted with the paddle attachment and mix on low speed. Do not cream heavily, just gently combine the butter and sugar. Add the egg and vanilla and mix until combined, scraping once or twice. Add the flour and salt all at once. Mix on low speed just until the dough comes together. Gather the dough into two 1-inch-thick disks. Wrap the disks in plastic and chill for 4 hours or overnight.

- Remove the tart dough from the fridge. If you are making the 9-inch tart, you will need only to use one of the disks. However, if you are making the six 4-inch tarts, you will need to use both disks of dough. The dough will be quite firm. Let it sit at room temperature for 5 to 15 minutes. Place your tart pan(s) within close reach while you are waiting. Once the dough has a bit of give when pressed, it's ready to roll. (See How to Roll Pastry [Pie] Dough, page 13.)

- Once your dough is rolled out, immediately begin cutting the dough for your individual tarts. If you have a round cutter that is ½ to 1 inch larger than your tart pan, that will be helpful. Otherwise, use a pastry wheel or knife to cut out circles that are at least ½ inch larger than the top diameter of your tart pans. Place the dough circles on the tart pans, and begin gently working them into the pans, making sure they are sitting well into the corners of the pans. Gently secure the dough against the side (don't press too hard). With a paring knife, trim extra dough from the top of each tart pan. Prick the bottom and sides of the tart dough several times with a fork, and place the pans on a baking sheet. Freeze the tart shells for at least 30 minutes. Meanwhile, preheat the oven to 375 degrees F (190 degrees C).

- Remove the shells from the freezer, and bake them for 20 to 25 minutes or until they are golden brown and firm. (Well-baked tart shells will have a deep golden color and be crisp.) Let them cool for 10 minutes before you pop the tart shells out of their pans and move them to a parchment-lined baking sheet.

Note: Dough that has been rolled out once can be reused. Just shake off the flour and press the scraps together or into another piece of dough. Triple wrap the disk in plastic wrap and hold in the refrigerator for 5 days or in the freezer up to 2 months. Thaw frozen dough in the refrigerator overnight.

^ Raspberry Custard Tart > Dark Chocolate Cake with Raspberry Orange Compote

Dark Chocolate Cake with Raspberry-Orange Compote

Makes 10 to 12 servings

For the cake:
2 cups (250 g) fresh raspberries

¾ cup (150 g) plus 2 tablespoons granulated sugar, divided

2 tablespoons water

2⅓ cups (303 g) all-purpose flour

⅓ cup (35 g) cocoa powder

1 teaspoon kosher salt

1 teaspoon baking soda

1 cup (2 sticks; 227 g) unsalted butter, room temperature

1 cup (218 g) brown sugar

3 eggs, room temperature

4 ounces (113 g) unsweetened chocolate, melted

¾ cup (175 mL) plain Greek yogurt or sour cream

For the compote:
1 large orange

3 tablespoons brown sugar

2 tablespoons honey

1 teaspoon orange zest

¼ teaspoon kosher salt

4 cups (500 g) fresh raspberries

2 tablespoons Grand Marnier or other orange liqueur

This recipe is sure to please chocolate lovers. This is a European-style gâteau—just one tall layer—super moist and with a beautiful crumb. Sifting aerates the flour, and creaming butter and sugar until light and fluffy imparts tiny bubbles into the mixture. Together, these techniques help create a light, uniform cake crumb. This cake also pairs nicely with the White Chocolate Mousse with Poached Sweet Cherries (page 46).

1. *To make the cake,* combine the raspberries, 2 tablespoons of the granulated sugar, and water in a medium saucepan and bring to a boil. Reduce the heat to a simmer, and cook the raspberries until they break down into a sauce. Push the raspberry pulp and juice through a fine-mesh strainer. Cool the puree to room temperature or chill before using.

2. Preheat the oven to 350 degrees F (175 degrees C). Lightly grease the bottom of a round 10-inch cake pan and line with a circle of parchment paper. Do not grease the sides of the pan. (This cake releases easily from the sides, and greasing the sides means the cake doesn't have anything to hold on to while it rises during baking.)

3. Sift the flour, cocoa powder, salt, and baking soda in a medium bowl.

4. In the bowl of a stand mixer fitted with the paddle attachment, cream the butter, brown sugar, and ¾ cup of granulated sugar on medium speed until light and fluffy (about 5 minutes), scraping once. Add the eggs one at a time, scraping well after each addition. Mix in the melted chocolate.

5. Combine the yogurt and raspberry puree in a medium bowl and set aside. With the mixer on low speed, mix half the dry ingredients into the butter mixture until almost combined. With the mixer running, add the raspberry mixture. Then mix in the remaining dry ingredients until almost combined. Scrape the bowl again and mix an additional 30 seconds on low speed until the batter is smooth.

6. Spoon the batter into the prepared pan. Bake the cake for 40 to 45 minutes or until the top springs back when touched. Cool to barely warm before removing the cake from the pan. Cool the cake completely before serving.

7. *To make the compote*, peel the orange and clean off the white pith and webbing. Cut the segments into ½-inch pieces (you should have about 1 cup, or 120 g). Combine the orange pieces, brown sugar, honey, zest, and salt in a medium pot and bring to a simmer over medium-high heat. Reduce the heat to medium-low and continue to simmer, about 5 minutes. Stir in the raspberries and the Grand Marnier. Cover and refrigerate the compote overnight.

8. If you've made the compote the day before, about 2 hours before serving, remove the compote from the refrigerator. Cut the cake into wedges, and serve the compote alongside the cake. Hold leftover cake covered at room temperature and the compote in the refrigerator for up to 2 days.

BLUEBERRIES

A cooked blueberry barely resembles its fresh predecessor, and yet both are revered. I grew up picking blue huckleberries. On family camping trips in huckleberry country, Mom would send us out with a can to pick some huckleberries for morning pancakes. As much as she liked huckleberries, I now wonder if it was also an excuse to get us out from underfoot. As kids, we also picked a lot of blueberries at a farm along the Sammamish River. With recycled paint pails tied around our waists with twine, we easily stripped the branches, hearing the blueberries plop, plop, plop into our pails. The picking was fast, and we were soon home washing and freezing them for pie later in the year.

There are many varieties of blueberries here in Washington, such as Elliott, Last Call, Titan, and Vernon. I tend to covet the petite varieties—maybe because they look and taste more like huckleberries. Varieties ripen throughout the summer, allowing the season to extend longer than just a few weeks in July. Possibly the best characteristic of blueberries is the ease with which they are washed and frozen, so you'd be smart to freeze as many as possible during the season.

Classic Lattice Blueberry Pie

I think blueberry pie is a truly American dessert, maybe because blueberry tends to satisfy most palates. Remember that berry pies need to cool completely before serving in order to set the filling. I like my pie to slump slightly when it's cut. If it stands up straight, I immediately think, "Too much thickener." Great berry pies are full of berries and just enough thickener to hold them together.

Makes 6 servings

1 recipe Flaky Pie Dough (page 12)

7 cups (about 1.25 kg) blueberries, fresh or frozen

¾ cup (150 g) sugar, plus extra for finishing

⅓ cup (74 g) cornstarch (see note, page 69)

½ vanilla bean, split and scraped, or ½ teaspoon ground vanilla bean, or 1 teaspoon vanilla bean paste

¼ teaspoon kosher salt

1½ tablespoons unsalted butter

2 tablespoons whole milk, for finishing

Vanilla ice cream (optional)

1. Prepare the pie dough.

2. Combine the berries, sugar, cornstarch, vanilla bean pulp, and salt in a large bowl. Toss until the vanilla is distributed and all the fruit is coated with sugar and cornstarch.

3. Roll out the bottom crust to about ⅛ inch thick and line a 9-inch pie pan. (See How to Roll Pastry [Pie] Dough, page 13.) Trim the edge of the bottom crust with scissors or a knife to ¼ inch over the edge of the pie pan. Toss the blueberry filling a few more times to pick up sugar sitting on the bottom of the bowl, and spoon the blueberry filling into the pie pan, mounding the berries 1 to 2 inches above the edge of the pie pan. Break up the butter in nickel-size pieces and distribute over the top of the berries. (If you have extra sugar and cornstarch at the bottom of the bowl, just sprinkle it over the top of the berries in the pie pan.) →

4. To construct the lattice crust, roll out the top crust to no less than ⅛ inch thick. Using a pastry wheel or knife, quickly cut the top crust into ¾-inch strips. Starting on the left side of the pie, place one of the shorter strips just inside the edge of the pie over the filling. Now take the next strip of dough and place it perpendicular to the first strip and across the top end of the pie. Now the trick is to continue going back and forth, one strip down, one strip across as you move across the pie, leaving about a ¼-inch space between strips so that the blueberry filling shows through. To get a nice weave, you will need to lift the strips that are running *under* the previous strips (that's half the strips, or every other strip), and fold them back to the left or top before placing the next strip. Then unfold the strips over the new strip, and do the same across the top of the pie. Continue this process until all the pie is covered with lattice.

5. Trim the lattice edges to ¼ inch over the edge of the pie plate. Pick up the overhanging edges, and turn them under all the way around the pie, pressing the ends of each strip of dough onto the rim of the pie pan, or just inside the pie pan if your pan doesn't have a flat edge. Crimp the edge with a fork or between your fingers to create a seal and a pretty edge.

6. Chill the finished pie in the freezer for 30 minutes before baking. While the pie is chilling, preheat the oven to 400 degrees F (200 degrees C).

7. The pie crust should be firm before going into the oven. (Freezing the pie keeps the dough from melting before it bakes and helps achieve a baked crust on the bottom.) Before baking, brush the top of the pie with milk and sprinkle sugar over the top, if desired.

8. Bake the pie for 20 minutes or until the top of the pie is dry and just starting to color. Reduce the oven temperature to 350 degrees F (175 degrees C) and continue baking for 30 to 40 minutes more, or until the pie is brown and bubbling. The pie should be golden on top, and you should be able to hear or see the juices bubbling in the middle of the pie. It's OK if some of the fruit juices bubble out of the pie.

9. It's especially important to cool a berry pie to room temperature before serving. A little vanilla ice cream doesn't hurt either.

Note: Flour or cornstarch for thickening the fruit filling? The biggest difference in results is that flour-thickened juices run opaque (a.k.a. cloudy), while cornstarch juices run nearly clear. I always use flour for apple pies, but cornstarch is a better choice for berry and cherry pies.

Blueberry-Cinnamon Crepes

It may be hard to imagine cinnamon and blueberries together. But somewhere back in time I must have been divinely inspired, because this combination works and has been a favorite since I created it years ago. Be sure to fold the crepes the French way, in half and then in half again, with the first-cooked side showing on the outside. At least that's how Jacques Pépin does it.

Makes 6 servings

1 recipe Crepes (recipe follows)

4 cups (680 g) blueberries, fresh or frozen

½ cup (100 g) sugar

¼ cup (57 mL) plus 2 tablespoons water, divided

2 teaspoons freshly squeezed lemon juice

¾ teaspoon ground cinnamon

2 tablespoons cornstarch

¼ cup (29 g) powdered sugar (optional)

1 teaspoon lemon zest (optional)

1. Prepare the crepes. Hold at room temperature.

2. Combine the blueberries, sugar, ¼ cup (57 mL) of the water, lemon juice, and cinnamon in a medium pot. Bring slowly to a simmer over medium heat. Reduce the heat to medium-low and simmer gently for 20 minutes. (Cooking the filling too quickly will cause the berries to fall apart.)

3. Make a slurry (thickening agent) by combining the cornstarch and the remaining 2 tablespoons of water in a small bowl until smooth. Add the slurry to the blueberries and stir to combine. Continue to simmer the blueberries for 1 minute to thicken the compote. Remove the compote immediately from the heat and pour into a bowl to cool. Chill the berry compote for 1 hour or until cold. The compote can be made up to 2 days ahead.

4. Place a crepe on a dessert plate. Spoon about ⅓ cup (about 100 g) of compote onto the crepe. Fold the crepe in half and then in half again so that it resembles a triangle. Move the first crepe to the side, and repeat with a second crepe and compote on the same plate. Repeat for all six plates. If you have leftover compote, drizzle some over the top. If using the lemon zest and powdered sugar, sprinkle a little powdered sugar and zest over each serving. Serve immediately.

CREPES

Makes about twelve 8-inch crepes

1¼ cups (163 g) all-purpose flour
¼ cup (25 g) sugar
1 teaspoon kosher salt
7 tablespoons unsalted butter, divided
1½ cups (340 mL) whole milk
3 eggs, room temperature
3 egg yolks, room temperature

- Whisk together the flour, sugar, and salt in a large bowl.

- Combine 6 tablespoons of the butter and the milk in a small saucepan over medium heat until just warm to the touch and the butter has melted. (If it gets too warm, transfer it to a metal bowl to cool to lukewarm.)

- In another bowl, whisk together the eggs and egg yolks. Add the eggs to the dry mixture and whisk until combined. The batter will become a thick paste. Add the milk and butter all at once and whisk gently until combined. Let the crepe batter sit for 2 hours or refrigerated overnight.

- When you are ready to make the crepes, cut twelve pieces of parchment about 6 inches square. These will separate your finished crepes and keep them from sticking.

- Heat a heavy 8-inch skillet or crepe pan over medium-high heat. Once the pan is hot, add 1 tablespoon of butter and swirl the pan to coat. Stir the crepe batter well before starting to cook the crepes. Add ⅓ cup of the batter to the pan and quickly tilt it to distribute it over the entire bottom of the pan. Return the pan immediately to the heat. Cook the crepe until the top bubbles and dries and the edges start to curl, about 1 minute. The underside of the crepe should be lightly browned. Using a heat-resistant rubber spatula, gently loosen the edges of the crepe. Slide the spatula under the crepe and flip it over. (I like to lift the edge of the crepe with my fingers, slide the rubber spatula under, and flip it over. You can use your

fingers and the spatula to straighten it in the pan if needed.) Cook the crepe on the second side for less than 1 minute, or until it is brown and releases easily from the pan. Only add butter for the first crepe. The butter will season the pan, and your pan will no longer need butter to cook each crepe. (If your crepes stick, the pan is probably not hot enough. Clean out your pan and start again.) Slide the cooked crepe onto a piece of the cut parchment. (Sometimes the first crepe is a "practice crepe" and not great for serving, but makes a good snack.)

- Continue cooking the crepes until all the batter is gone. It is important to stir the batter frequently, as the solids will settle toward the bottom. You can make as many crepes as you need and store the remaining batter in the refrigerator for up to 2 days.

^ Blueberry-Cinnamon Crepes > Blueberry and Lemon Curd Tart

Blueberry and Lemon Curd Tart

Blueberries and lemons are a common pairing, but here I use gently spiced poached blueberries to add another dimension to a familiar tart. The poached blueberries both contrast with and complement the bright, acidic lemon curd. Meyer lemons (a hybrid of a citron and a mandarin orange / pomelo) work wonderfully here, so try making this in winter when they are in season.

Makes 8 to 10 servings

One 9-inch fully baked sweet tart shell (page 58)

1 recipe Poached Blueberries with Vanilla Bean and Anise (page 79)

For the filling:
1 cup (200 g) granulated sugar, divided

¼ cup (34 g) cornstarch

⅛ teaspoon kosher salt

½ cup (113 mL) freshly squeezed lemon juice (from 4 to 5 lemons)

6 egg yolks

4 eggs

2 teaspoons grated lemon zest

½ cup (113 g) unsalted butter

1. Prepare the tart shell, then the poached blueberries.

2. *To make the filling*, whisk together ½ cup (100 g) of the sugar, cornstarch, and salt in a medium bowl. Whisk 2 tablespoons of the lemon juice into the dry mixture. In another medium bowl, whisk together the egg yolks and eggs.

3. Stir together the remaining ½ cup (100 g) of sugar, the remaining lemon juice, and zest in a medium pot, and bring it to a boil over high heat.

4. Have a clean bowl and spatula ready by the stove for your finished curd before you begin cooking it. Quickly whisk the sugar and cornstarch mixture into the eggs until well combined. Temper the eggs by slowly pouring the hot liquid into the egg mixture while whisking. Then return the entire mixture to the pot while whisking. Cook over medium heat, whisking constantly, until the curd

thickens and boils. Continue cooking, whisking quickly, until the curd boils for 1 minute. Remove the curd from the heat, and stir in the butter. Pour the curd into a bowl and refrigerate until cold, about 2 hours.

5. When both the tart shell and curd have cooled, spread the curd in the shell. Chill the lemon-filled tart in the freezer until cold (about 1 hour) and then move it to the refrigerator for another hour.

6. While the tart is being chilled, drain the poached blueberries well. (Let them sit in a strainer or colander for 5 minutes, shaking them occasionally to get rid of moisture.) Remove the tart from the refrigerator, and gently spoon the berries over the top of the lemon filling, covering the filling. It's best to chill the tart for another hour before serving. Keep the tart refrigerated and loosely covered for up to 3 days.

Poached Blueberries with Vanilla Bean and Anise

These infused blueberries hold for several weeks refrigerated and will brighten up desserts and breakfasts alike. I love to infuse fruit with flavor while still preserving the integrity of the fruit. This recipe does both. Be sure to remove the cinnamon stick and anise so they don't overpower the fruit while it's being held. I find that two hours is about right for infusing, but taste and decide for yourself. Enjoy these blueberries on their own, over yogurt or on pancakes, or with the White Chocolate Mousse on page 46.

Makes 3 cups

2 cups (454 mL) water

1¾ cups (375 g) sugar

1 cinnamon stick

2 star anise

½ vanilla bean pod

4 cups (680 g) blueberries, fresh

1. Combine the water, sugar, cinnamon stick, star anise, and vanilla bean pod in a large pot. Bring to a boil over high heat, stirring a few times. Place the blueberries in a large bowl, and pour the poaching liquid over them, then cool to room temperature. If they haven't poached enough for your taste (remember, they shouldn't be falling apart), return the berries and poaching liquid to the pot and bring back up to hot. Then remove from the heat and pour back into the bowl to cool.

2. Chill the berries in the poaching liquid until ready to use, at least 1 hour. They will keep for 4 weeks refrigerated.

BLACKBERRIES AND MARIONBERRIES

We get grumpy about blackberries in Washington because the Himalayan (non-native) variety grows everywhere and can quickly take over a garden. (I'm constantly trying to tame them on the edge of my yard.) Cooks favor the elusive, wild Cascade blackberries, with delicate vines that run along the forest floor and treasured petite berries prized for their flavor. (I've rarely seen them myself in the forest.) Along with sweet blackberries, I enjoy the tarter marionberry (essentially a hybrid of two blackberry varieties) and the earthy, dark red tayberry (a blackberry-raspberry hybrid).

Although hardier than raspberries, blackberries and their cousins require similar care after picking. Don't wash them until you're ready to use them—just lay them out on parchment-lined baking sheets in single layers with a piece of plastic wrap loosely over the top. You can pull from these for the next two to four days.

Marionberry Crostata with Whole Grain Crust

If you have only a small amount of time to put together a pie-like dessert, a crostata is for you. We call them open-faced tarts or rustic tarts in the United States. They use just one crust and about half the filling of a full pie. Although you could make this with a standard white flour crust, I encourage you to try the whole wheat pastry flour. The nutty, whole-grain crust will offset the intense marionberry filling and give this tart a "best of both worlds" feeling.

Makes 6 servings

1 disk from Whole Wheat Pie Dough recipe (page 12; see variation)

4 cups (580 g) marionberries (or blackberries, tayberries, or a combination)

⅓ cup (67 g) sugar, plus 2 tablespoons for finishing

¼ cup (34 g) cornstarch

½ vanilla bean, split and scraped, or ¼ teaspoon ground vanilla bean, or ½ teaspoon vanilla bean paste

⅛ teaspoon ground nutmeg

Pinch kosher salt

1 tablespoon unsalted butter

2 tablespoons whole milk, for finishing

1. Prepare the pie dough.
2. Prepare a baking sheet by lightly greasing the pan, lining with parchment, and then greasing the parchment. Inevitably, this tart leaks juices and will stick even to parchment.
3. Combine the marionberries, ⅓ cup (67 g) of the sugar, cornstarch, vanilla bean pulp, nutmeg, and salt in a large bowl and toss gently.
4. Roll out the pie dough to about 10 inches diameter and ⅛ inch thick. (See How to Roll Pastry [Pie] Dough, page 13.) Using a pastry wheel or knife, trim the perimeter of

the dough. Don't worry if the dough isn't round. Even an oval will make a fine-looking crostata.

5. Flip the dough over (with the more floured side up), and move it to the parchment-lined baking sheet. Gently stir the filling a few more times and spoon it into the middle of the dough, leaving a 2-inch border around the outside edge. Break up the butter into nickel-size pieces and distribute them over the top of the fruit.

6. Pleat the dough over the fruit, turning the tart and pleating as you move around the exterior. There will be about a 4-inch open circle of fruit in the middle. Freeze the crostata for 30 minutes or until the dough is solid; this helps the crostata hold its shape while baking. (The crostata could be frozen overnight and baked the next day.) While the crostata is in the freezer, preheat the oven to 400 degrees F (200 degrees C).

7. When you are ready to bake, brush the top of the crostata with milk and sprinkle with the remaining 2 tablespoons of sugar. Bake the crostata for 15 to 20 minutes, or until the crust is dry and starting to brown. Reduce the temperature to 350 degrees F (175 degrees C), and continue baking the crostata until it is dark golden brown and the juices are bubbling, 20 to 25 minutes more.

8. Cool the crostata to room temperature or barely warm before serving. Store leftovers in an airtight container at room temperature for 2 days or in the refrigerator for up to 5 days.

^ Marionberry Crostata with Whole Grain Crust > Pain Perdu with Assorted Berries and Grand Marnier

Pain Perdu with Assorted Berries and Grand Marnier

Pain perdu ("lost bread") has a long history, hailing back to medieval times, when it was necessary to use every bit of leftover food, including stale bread. Yes, it's also the predecessor of American French toast, but on a much grander, richer scale. In the days of tableside preparations, pain perdu was a giant in the dessert world, along with Crêpe Suzette. Although berries are the most traditional fruit accompaniment, most any other fruit will work with this classic dessert. Get all your ingredients ready ahead of time and organize your equipment so it's quick and easy to prepare when it's time for dessert.

Makes 4 servings

1 egg

3 tablespoons sugar

3 tablespoons unsalted butter, melted

½ vanilla bean, split and scraped

½ teaspoon ground cinnamon

⅛ teaspoon ground nutmeg

Pinch kosher salt

¼ cup (33 g) all-purpose flour

¾ cup (177 mL) whole milk

½ loaf (about 450 g) brioche, challah, or other egg bread (day-old bread is best)

3 tablespoons unsalted butter, divided

½ cup (113 g) Grand Marnier or other orange liqueur or brandy

2 cups (290 g) blackberries and assorted berries, such as raspberries, tayberries, marionberries, and blueberries

1. Whisk the egg in a medium bowl. Then whisk in the sugar, butter, vanilla bean pulp, cinnamon, nutmeg, and salt. Finally, whisk in the flour and then the milk. Let the batter stand at room temperature for at least 1 hour.

2. With a serrated knife, slice the bread into ½-inch-thick slices. Then cut the slices crosswise into triangles. Depending on the size of your bread loaf, you may want to use two or three triangle slices per serving.

3. Heat a large sauté pan over medium-high heat. While the pan is heating, whisk the batter again and pour it into a wide pan so that you can start to soak the bread slices in the batter. If your sauté pan is 10 inches in diameter, you should be able to cook four half slices at a time (or two servings), so add four slices of bread to the batter. Soak the slices for 30 seconds, then flip them over and soak another 30 seconds.

4. Melt 2 tablespoons of the butter in the hot pan. The pan should be hot but not smoking, and the butter should sizzle. (It's OK for the butter to brown—this adds flavor—but you don't want it to turn black.) Place the soaked pieces of bread in the pan and brown them well on both sides, 2 to 3 minutes per side. (Adjust the temperature as needed. Bread that browns too quickly won't be cooked inside; thus the interior ends up being a bit mushy.)

5. Immediately move the cooked bread to dessert plates, then soak and cook the second batch of bread slices.

6. After cooking the bread, turn the heat up to high and add the remaining 1 tablespoon butter to the hot pan and pour in the Grand Marnier. (You may want to light the spirits to burn off some of the alcohol, but it isn't necessary.)

7. Add the berries and quickly toss them to coat. Spoon the berries and sauce over the bread and serve immediately.

Bittersweet Chocolate Tart with Blackberries and Basil

This tart came to me during the heat of summer, when basil was overflowing the buckets of one stand at the farmers' market, and right next door was the berry farm stand stacked high with flats of blackberries. I bought some of both and went home to create this distinctly summer tart.

Makes 8 servings

One 9-inch, fully baked sweet tart shell (page 58)

For the filling:
8 ounces (227 g) bittersweet chocolate (65% to 72% cacao)

3 tablespoons honey

1 cup (227 mL) heavy cream

3 tablespoons unsalted butter, room temperature

For the compote:
4 cups (580 g) blackberries

⅓ cup (67 g) sugar

2 tablespoons water

2 star anise

½ vanilla bean pod

4 tablespoons chopped fresh basil, plus extra for garnish

1. *To make the filling,* chop the chocolate into ½-inch pieces. In a medium bowl, combine the chocolate and honey. Heat the cream in a small pot over medium heat to very hot, and then pour it over the chocolate. Stir until the chocolate is melted. Stir in the butter until melted and the filling is shiny and smooth. (This filling is essentially a ganache; it should never look separated, curdled, or oily.)

2. Immediately pour the filling into the prepared shell, tapping gently to level it. Freeze the tart for 30 minutes to set the ganache, and then move it to room temperature. (The tart should not be eaten chilled or cold because the ganache will be too firm.)

3. *To make the compote,* combine the blackberries, sugar, water, star anise, and vanilla bean in a medium pot.

→

Cook over medium heat, stirring gently and occasionally. Bring to a gentle simmer and continue to cook for about 10 minutes, or until the berries have softened slightly. (Most of the berries should stay intact with some broken down to make sauce.) Remove the compote from the heat, and use a slotted spoon to transfer just the berries to a bowl. Continue simmering the sauce left in the pot for 2 to 3 minutes over medium heat, stirring frequently until it thickens slightly. Pour the sauce over the berries, stir in the basil, and cool the compote to room temperature, at least 30 minutes.

4. When you are ready to serve the tart, cut it into eight pieces and place them on dessert plates. Spoon the berry compote onto each piece of the tart. Sprinkle with additional chopped basil and serve immediately. The tart will hold refrigerated without the sauce for up to 5 days.

Vanilla Bean Cake with Glazed Blackberries and Stone Fruits

Every fruit-dessert book needs a grand cake, and this is a beaut. What I especially love about this cake is how easy it is to put together. Use blackberries and any other stone fruit you can find at the time. You will get plenty of oohs and aahs serving this at the table. For a completely adult dessert, add some apricot brandy to the apricot glaze.

Makes 8 to 12 servings

1 recipe Classic White Cake (recipe follows)

For the blackberry and stone fruit glaze:
2 cups (454 g) apricot jam

¾ cup (170 g) water

3 cups (430 g) blackberries

3 cups (430 g) sliced stone fruit, such as apricots, nectarines, or plums

1. Prepare the cake batter and bake the cakes.

2. *To make the blackberry and stone fruit glaze*, combine the apricot jam and water in a medium pot. Bring to a simmer over medium heat, and stir until the jam is liquid. Strain the glaze through a fine-mesh sieve and cool to room temperature before using.

3. Pick your best berries and fruit slices, and combine them in a large bowl. Pour half the apricot glaze over the fruit, and gently fold the berries and fruit so that they are covered in glaze. (If you have particularly fragile fruit, you can dip four to five pieces at a time in the reserved glaze.) With a slotted spoon, scoop the fruit out of the glaze and place it in a single layer on a parchment-lined baking sheet. You could do this step a few hours →

ahead, and then refrigerate the fruit until you are ready to build the cake. (The glaze will keep the sliced fruit from oxidizing.)

4. Using a serrated cake knife, remove the tops of the cakes to make a flat surface and discard the tops. Place the bottom layer on a cake plate, and spread about ¼ cup (57 g) apricot glaze on the first layer. Top with about 2 cups (about 280 g) of the assorted glazed fruit. (Let the fruit peek out from the side of the cake.) Place the top layer on the cake and spread with another ¼ cup of the apricot glaze.

5. Pile the remaining glazed fruit on top of the cake, mounding some in the middle. Generously brush more apricot glaze on top of the cake. (Feel free to drizzle some down the side, too.) If the glaze is setting up while you are working with it, warm it gently to loosen.

6. Chill the cake for 30 minutes before serving. Serve close to room temperature for the best flavor.

CLASSIC WHITE CAKE

Makes two 8- or 9-inch cakes

2½ cups (325 g) cake flour
2 teaspoons baking powder
½ teaspoon kosher salt
2 cups (454 g) sugar
1 cup (2 sticks; 227 g) unsalted butter, at room temperature
4 eggs, room temperature
1 cup (227 mL) buttermilk
2 teaspoons vanilla extract

1. Preheat the oven to 350 degrees F (175 degrees C). Lightly grease the bottoms of the cake pans, and line them with circles of parchment paper. (Do not grease the sides of the pans.)

2. Sift together the flour, baking powder, and salt in a medium bowl.

3. Using a stand mixer fitted with the paddle attachment, cream the sugar and butter until light and fluffy, about 5 minutes, scraping once. Add one egg at a time, mixing on medium speed until incorporated and scraping after each addition.

4. Combine the buttermilk and vanilla in a separate bowl. With the mixer on low speed, add half the buttermilk mixture to the mixer bowl, then half the flour mixture, then the remaining buttermilk mixture, and finally, the remaining flour mixture, in one continuous series. Mix until most of the flour is incorporated. Stop and scrape well and then mix for another 10 to 20 seconds, or until the batter is smooth. Be careful not to overmix once you begin adding the flour. Overmixing will make the cake tough.

5. Divide the batter between the cake pans. Bake the cakes until golden and the tops spring back when touched, 35 to 40 minutes.

6. Let the cakes cool to barely warm in the pans on a wire rack. Unmold them by running a small knife around the edge of the pans and then turning out each cake onto a rack or cake plate.

7. The cakes can be frozen (triple wrapped in plastic wrap) for up to 1 month.

APRICOTS

Although perfectly palatable and simple to eat, fresh apricots just don't do it for me. But cooked? Wow! Whether roasting, grilling, stewing, or poaching them, cooked apricots radiate flavor. An added bonus is the simplicity with which they can be readied for cooking because there is no peeling involved. (Although, they are fragile. Sometimes I can barely get them home before they are bruised and soft.)

Common varieties of apricots are Perfection, Goldbar, and Rival. Some are completely gold; others have a red blush. Like my mother, I tend to favor the Perfection apricot for making jam (and for the Apricot and Walnut Streusel Bars, page 102). Overripe apricots are perfect for jam making, and they can be quickly cooked down into a puree for freezing and storing. Firm apricots (just slightly underripe) are important if you're going to grill or poach them.

Grilled Apricots with Brown Butter and Maple-Tamari Glaze

Step out onto the deck and try your hand at grilling fruit. Grilled apricots pair well with the brown butter, which elevates this dessert. Tamari is generally a gluten-free alternative to soy sauce and is also thicker and less salty. This is possibly the most savory recipe in this book, with its umami overtones, and would be a good addition to fish and meat dishes.

Makes 6 servings

6 tablespoons unsalted butter

2 tablespoons maple syrup

2 tablespoons tamari or soy sauce

¼ teaspoon ground cinnamon

¼ teaspoon ground ginger

Pinch kosher salt

6 large apricots (just ripe and firm)

2 tablespoons canola oil (or other high-heat oil)

1. Melt the butter in a small saucepan over medium-high heat, letting it foam and then brown. You should see brown bits in the bottom of the pan (but not black). The butter should smell pleasantly nutty. Remove the pan from the heat, and pour the butter into a small bowl. Add the maple syrup, tamari, cinnamon, ginger, and salt and whisk to combine.

2. Split the apricots from end to end into halves, removing the pits. Lay out the apricot halves cut side down on a baking sheet or plate. Preheat the grill. If you are using a gas grill, set the heat on medium flame. (If your grill tends to run hot, the lowest heat will work fine.) Charcoal grills can be more finicky, but with either grill, work with medium-low heat. (You could also prepare these with a

grill pan on the stove.) Pour the oil on a folded paper towel, and use tongs to wipe the lightly oiled paper towel down the grill grate. Use the tongs to place the apricot halves cut side down on the hot grill in a single layer, at least 1 inch apart. Grill for 3 to 5 minutes, or until there are distinct grill marks on the bottom of the apricots and they start to soften. Carefully turn over the apricots with the tongs. Generously brush the glaze on the apricots and grill another 2 minutes, or until they are cooked through and tender but not falling apart.

3. Remove the apricots from the grill, and place them cut side up on a serving plate. Heat the remaining glaze in a small pot, whisking well, and drizzle over the apricots. The apricots can be eaten warm or at room temperature, and extras are wonderful the next morning on your yogurt.

< Grilled Apricots with Brown Butter and Maple-Tamari Glaze ^ Apricot and Walnut Streusel Bars

Apricot and Walnut Streusel Bars

Every summer, I inevitably end up with overripe apricots. The only reasonable way to deal with them is to stew them and freeze the sauce, or make this recipe. These bars are more delicate than similar versions because this quick homemade jam contains less sugar (and more intense flavor) than store-bought jam. The nuttiness of the whole wheat flour and walnuts along with the candied ginger makes a flavorful crust, and a perfect contrast to the tart apricot filling.

Makes about 24 bars

For the jam:
4 cups (about 1 kg) quartered apricots

1 cup (200 g) sugar

¼ cup (57 mL) water

¼ teaspoon kosher salt

For the bars:
1 cup (2 sticks; 227 g) unsalted butter, room temperature

2 cups (270 g) whole wheat flour

1 cup (213 g) brown sugar

1 teaspoon ground ginger

½ teaspoon kosher salt

1½ cups (170 g) chopped walnuts (½-inch pieces)

2 tablespoons candied ginger

1. Preheat the oven to 375 degrees F (190 degrees C).

2. *To make the jam*, combine the apricots, sugar, water, and salt in a medium pot over medium heat and bring to a simmer. Reduce the heat to keep the apricots simmering and cook for 30 minutes. Transfer the jam to a metal bowl and cool to room temperature before using. (The jam can be made 1 day ahead and refrigerated.)

3. Prepare a 9-by-13-inch pan. Lightly grease the bottom and sides of the pan. Cut a piece of parchment to set in the pan as a sling, running the long ends of the parchment up the sides of the pan. The other two sides of the pan will not be covered. This gives you an easy way to remove the bars without having the parchment gathered in the corners of the pan.

4. *To make the bars*, cut the butter into eight pieces. Combine the flour, butter, brown sugar, ground ginger, and salt. Mix with your hands or on low speed with a mixer fitted with a paddle attachment until combined and the dough is moist and holds together when pressed. Add the walnuts and candied ginger and mix to combine.

5. Press half the dough into the bottom of the pan. Place another piece of parchment on top of the dough, and use your hand to press and smooth the dough. Remove the top parchment.

6. Bake the bottom crust for 20 to 25 minutes until lightly browned and firm. Remove the crust from the oven, and spread the cooled apricot jam over the top. Sprinkle the remaining dough as a streusel over the jam. There will be a lot of topping, but it will settle and bake, compacting some.

7. Bake the bars for 35 to 40 minutes until the top crust is brown and the filling bubbles across the top of the bars.

8. Cool the bars in the pan for 20 minutes. Then use a knife to gently loosen the bars from the two sides of the pan without parchment. Use the parchment ends to gently pick up the bars, and set them on a rack to cool completely before cutting into 1½-inch squares. You will get the cleanest cuts if you cool the bars to room temperature and then freeze them until firm, about 1 hour. Store well wrapped in the refrigerator or freezer.

Poached Apricots in Sauternes

Poaching is a favorite fruit-cooking technique that easily and gently infuses flavor into the fruit. Although a French Sauternes is my preference for the poaching liquid, another sweet white wine would also work well. The addition of bay leaf turns a classic poach into something earthier, almost herbal. Once done with the poaching liquid, refrigerate it for your next poach or reduce it to a syrup and use it on fresh fruit or salad greens.

Makes 6 servings

3¼ cups (750 mL) Sauternes or other sweet white wine, such as gewürztraminer

3 cups (687 mL) water

½ cup (100 g) sugar

¼ cup (57 mL) mild honey

2 strips orange zest

2 strips lemon zest

2 bay leaves

2 star anise

1 vanilla bean (see note, page 106)

8 firm apricots

1. Combine all ingredients except for the apricots in a large pot and bring to a boil over high heat, stirring a few times while it heats. Reduce the heat to the lowest setting for about 30 minutes so that the flavors meld. (Longer is better—you could let it steep for up to 2 hours.)

2. When the poaching liquid is ready, carefully halve and pit the apricots so the halves are clean and pretty. Pour the poaching liquid into a large bowl, and add the apricot halves into the poaching liquid, flipping them so they are all cut side up. Cut a circle of parchment paper about 2 inches larger than the diameter of the pot. Place the parchment on top of the liquid, pressing it to the surface. (If your apricots are very firm or slightly unripe, you may want to keep them in the pot and on the heat for 10 minutes before putting them in the bowl.) →

3. Poach the apricots until they are tender, flipping them once. They should have some give when touched, but not be falling apart. Depending on the ripeness of the apricots, this can take from 20 to 30 minutes.

4. Store the poached apricots (in less poaching liquid if you wish) refrigerated until they are cool. Be sure to keep the parchment on the surface to keep the apricots from browning. Poached fruit can be kept for about a week in the poaching liquid, which will deepen the flavor. In this case, you may want to remove the anise and bay leaf once you achieve the preferred infusion. (The poaching liquid can be reused three to four times before discarding.)

5. Serve the apricots cold or at room temperature. Make a sauce for the apricots by reducing some of the poaching liquid to a light syrup. Place two apricot halves in a small bowl or on a plate, and drizzle the syrup over them.

Note: I use the vanilla bean whole here so that I don't get vanilla bean pulp all over the beautifully poached apricots.

PEACHES AND NECTARINES

My mom's quest for getting our own fruit (whether picking it ourselves or buying it directly at the orchard) was adventurous and sometimes frustrating. Somewhere along the years, Mom talked Dad into adding a two-hour drive to our annual Mount Rainier family camping trip over the Cascades to sunny eastern Washington, where peaches and nectarines (and pears) grow abundantly. We'd load into the station wagon and trek over the mountains to search out the orchard my mom had picked that year. She'd communicate the sketchy directions in the annual orchard U-pick guide, and Dad would patiently follow them. Inevitably, we'd get lost. *Every single year.* No signs, just crossroads on the unending grid of farms. Dad would grumble, "They just take you out here and just *drop* you." Eventually, we'd find the orchard (or another orchard because we'd given up on the first one). We'd pick our own or pull from the giant bins of fruit. While Mom finagled the price with the orchard owner, the rest of us would stand in the orchard, happily slurping down peaches. Mom canned and froze a lot of peaches, and I remember it as a chore. But later in the winter, we enjoyed the literal fruits of our labors.

Peach varieties differ—some are better for canning, some for frozen, and some are perfect fresh. Luckily, peach and nectarine season is long, with some peaches available into late September or even early October. Some of my favorite varieties are Early Redhaven, Regina, Hale, and O'Henry. The long list of varieties means they ripen at different times throughout late summer, so get the list from your local orchardist and taste them all. You might even decide to take a similar orchard quest and pick your own.

Peaches and Nectarines Sautéed with Thyme

This is an all-time favorite recipe of mine, and it exhibits the essence of this book. Fresh peaches and nectarines are paired with brown butter, just a bit of sweetener and vanilla, and then quickly sautéed and finished with fresh thyme. The key word here is *quickly* as the fruit will fall apart with too much cooking. This same technique can be used for plums, apples, pears, or citrus. I love this simple sauté all on its own as a showcase of stone fruit, in particular. To step up this dessert, pair it with the Classic White Cake (page 94).

Makes 4 servings

2 large peaches

2 large nectarines

3 tablespoons unsalted butter

2 tablespoons light brown sugar

1 tablespoon honey

½ vanilla bean, split and scraped

1 teaspoon chopped fresh thyme

1. Cut the peaches and nectarines into ¼-inch-thick slices. Gather all the other ingredients before you start to cook.

2. In a 9-inch skillet over medium-high heat, add the butter, letting it foam and then recede, leaving bits of brown on the bottom of the pan and giving off a nutty aroma. (Don't take this step too far, or you will have black butter.)

3. Add the sugar, honey, and vanilla bean pulp, and cook until the sugar is dissolved, about 2 minutes. Increase the heat to high, and add the fruit slices. Sauté the fruit quickly, tossing three to four times. Add the thyme and toss to coat the fruit. Serve immediately.

^ Peaches and Nectarines Sautéed with Thyme > Peach Mousse

Peach Mousse

Fruit mousses are inherently French and not all that common in home kitchens. This is surprising, considering how easy they are to make. They perfectly showcase the true flavor of the fruit and, as in this recipe, they need no help from other flavors. Just pure peach. Pick your favorite peaches for this mousse, and be sure they are ripe and flavorful.

Makes 6 servings

9 medium yellow peaches (very ripe)

1 tablespoon sugar

1 tablespoon water

1 cup plus 2 tablespoons (266 mL) heavy cream

4 teaspoons powdered gelatin (measured from 2 packets)

3 egg whites

⅔ cup (66 g) sugar

1. Peel and chop eight of the peaches, and combine them with the sugar and water in a large pot. Cook on medium heat until the peaches are soft, about 10 minutes. Process the peaches (and all juice) in a blender until smooth. Strain the peach puree through a fine-mesh sieve or cheesecloth. Reserve 2 cups (454 g) of the puree for the mousse, plus a bit more for garnish if desired, and chill it until cold, about 1 hour. Any remaining puree can be frozen for several months.

2. Whip the heavy cream to medium-stiff peaks and refrigerate. Combine the gelatin and 3 tablespoons of the peach puree in a small stainless-steel bowl. Let sit for at least 5 minutes.

3. Make a Swiss meringue by starting a hot-water bath with 2 inches of water in a medium pot brought to a boil and turned to medium-low heat. Combine the egg whites and sugar in the bowl of a stand mixer, and whisk the mixture by hand over the water bath until it reaches 140 degrees F (60 degrees C). (The bowl should not touch the water.) Move the bowl to the stand mixer

fitted with the whisk attachment, and whip the egg mixture until the bottom of the bowl no longer feels warm.

4. While the egg whites are whipping, place the gelatin mixture over the water bath and stir until the mixture is just warm and the gelatin is melted. Temper some of the remaining peach puree into the gelatin mixture, whisking well, and then pour all the gelatin mixture into the rest of the puree and whisk well.

5. When the meringue is cool, remove it from the mixer and gently fold one-third into the peach puree. Then fold in the remaining meringue until the mixture is well combined. Temper the cream into the mousse by first folding in one-third of the cream. Then fold in the rest of the cream until the mousse is smooth and uniform.

6. Divide the mousse among six 8-ounce glasses or dessert dishes. Chill the mousse for at least 1 hour. If you make the mousse 1 day ahead, portion it and stretch a small piece of plastic wrap over each glass and refrigerate.

7. Before serving, finish the mousse with fresh peaches and/or some of the leftover peach puree. Serve immediately.

Roasted Peach Bread Pudding

I've been making this bread pudding since my Salish Lodge days, when I featured it with rhubarb. Peach is a winner here because of its tart-sweet flavor and its ability to meld well with the rich custard, which is essentially a crème brûlée base. Bread puddings enjoyed a well-deserved revitalization some years ago as the perfect comfort-food dessert. The Ginger Crème Anglaise finishes it perfectly. For a somewhat lighter version, replace the heavy cream with half-and-half in the custard.

Makes 6 to 8 servings

1 recipe Ginger Crème Anglaise (recipe follows)

For the custard:
4 egg yolks

2 eggs

2¼ cups (511 mL) heavy cream

½ cup (113 mL) whole milk

⅓ cup (67 g) sugar

2 teaspoons vanilla extract, or ½ vanilla bean, split and scraped, or ¼ teaspoon ground vanilla bean, or ½ teaspoon vanilla paste

For the roasted peaches:
3 medium peaches (firm), cut into 1-inch pieces, skin on

¼ cup (53 g) brown sugar

⅛ teaspoon salt

For the bread pudding:
6 cups (about 454 g) 1-inch cubes brioche or challah bread (day-old works well and some or all the crust may be trimmed off or left on)

Unsalted butter, for greasing

1. Prepare the crème anglaise.

2. *To make the custard,* set up a large metal bowl and fine-mesh sieve or sieve with four-ply cheesecloth. Whisk the yolks and eggs in a medium bowl. Combine the cream, milk, sugar, and vanilla bean pulp in a medium pot and cook over high heat until hot but not boiling. (If the cream boils, pour it into a metal bowl and let it cool for 10 minutes. Then continue with the next step from the bowl instead of the pot.) Temper the eggs by slowly adding the hot cream mixture to the eggs while whisking continuously. Strain the mixture through the sieve. Cool the custard to room temperature before using. The custard can be refrigerated up to 5 days.

3. *To roast the peaches,* preheat the oven to 350 degrees F (175 degrees C). In a large bowl, toss the fruit with the sugar and salt, and transfer the fruit to a parchment-lined baking sheet in a single layer. Roast the fruit until it is just tender to the touch but still holds together, 12 to 15 minutes. Cool the fruit to room temperature before adding it to the bread pudding.

4. *To make the bread pudding,* preheat the oven to 300 degrees F (150 degrees C). Combine the roasted peaches and bread cubes in a large bowl. Pour the custard over the bread cubes and gently toss to combine with your hands. Let the brioche sit to soak about 10 minutes or until the custard is absorbed into the bread. Generously butter a 9-by-13-inch pan. Gently move the bread pudding mixture to the prepared pan.

5. Bake the pudding for 35 to 40 minutes. The pudding will dry slightly on top and feel firm to the touch when it is done. Serve warm or at room temperature, topped with the crème anglaise. This dessert would also be wonderful at breakfast or brunch.

GINGER CRÈME ANGLAISE

Makes 6 servings

1 cup (227 mL) whole milk
1½-inch by ½-inch piece fresh ginger, peeled and cut into 6 slices
1 tablespoon light brown sugar
½ vanilla bean, split and scraped
4 egg yolks

- Combine the milk, ginger, sugar, and vanilla bean pulp and pod in a small pot. Set over medium-high heat until very hot. Remove from the heat, and let the ginger infuse into the hot milk for 30 minutes. Meanwhile, whisk the egg yolks in a medium bowl and set a fine-mesh sieve over a clean bowl.

- When the milk has been infused, reheat it until it's hot. Temper the egg yolks by pouring the hot milk in a slow stream into the yolks while whisking continuously. Return the mixture to the saucepan and cook over medium heat, stirring constantly with a heat-resistant spatula. Cook until the sauce thickens and coats the back of the spatula (nappé). Do not let the custard boil. Immediately remove the pan from the heat, and strain the sauce through the sieve.

- Chill the sauce in the refrigerator until cold, about 1 hour. The sauce will hold in the refrigerator for 5 days.

AUTUMN

PLUMS 123

Simple Stewed Plums and Pluots	124
Plum-Lavender Crisp	127
Fromage Blanc Cheesecake with Brown-Butter-Braised Plums	128

APPLES 133

Rosemary Apples in Crepes with Rum Caramel Sauce	134
Brandied Apple Pie with Ginger Streusel	138
Apple Upside-Down Cake with Spelt and Rye Flours	142
Cider-Braised Apples with Warm Winter Spices	144

PEARS 147

Pear and Fig Pie	148
Burgundy-Poached Pears with Mulled-Wine Syrup	152
Maple and Pear Panna Cotta	154

PLUMS

I love plums and the way they transform when cooked. Their skins (whether red or purple) color the entire dish, creating brilliant hues. A favorite plum is the Italian prune plum. It seems to get a less favorable rating in the grocery store because it's just not that pretty. But it is, by far, one of the most flavorful plums when cooked. I never think of plums without recalling a particular family incident at the US-Canada border in the late 1970s, when my mom sheepishly admitted to the border agent that we had fresh plums in our possession—a no-no for crossing into Canada. As my mom was thrifty, she insisted she was not giving up the plums. So my dad pulled our station wagon / travel trailer combo over to the side of the border entry and all seven of us consumed that big bag of plums in the shadow of the Peace Arch before we were allowed to drive into Canada. Although I don't remember the details, I suspect there were stomachaches later.

Plum varieties have broadened, and now we have not only plums but pluots, plumcots, and apriums. Any of these can be used as plums, although they all have different flavor profiles. I'm particularly fond of my own French prune plum tree, which produces a small golden oval plum with a red blush and a flavor that is almost floral. Other favorites are Blue Damson, Black Friar, and President plums. I'm also particularly taken with pluots such as Flavor Grenade and Dapple Dandy. The names don't hurt their appeal.

Simple Stewed Plums and Pluots

Stewing is the quickest and simplest cooking technique for fruit. These quickly stewed plums gain a little sweetness and flavor from sugar and spices. Cooked plums emit a depth and tartness not apparent in the fresh fruit. I recommend using a variety of plums and pluots of different colors. With the plum skin giving the stewed plums deep reds and purples, this dessert is beautiful served in a light-colored bowl.

Makes 6 servings

10 medium plums and pluots, halved and pitted

½ cup (113 mL) water

¼ cup (50 g) sugar

1 tablespoon freshly squeezed lemon juice

2 star anise

1 strip of orange or lemon zest, about 2 inches

½ cinnamon stick

½ whole nutmeg

¼ vanilla bean, split and scraped, or a pinch of ground vanilla bean

1. Combine all the ingredients in a large pot. Bring to a gentle simmer over medium heat, stirring occasionally. Once the sugar dissolves, reduce the heat and simmer slowly until the fruit is tender, stirring occasionally. Cook for at least 30 minutes to infuse the flavors; it may be necessary to reduce the temperature to the lowest setting so the fruit is barely simmering or even just steeping. Remove the spices and zest once the flavors are to your liking.

2. Serve the stewed plums and pluots with their juices when they are warm, at room temperature, or cold. The stewed fruit will keep for 1 week in the refrigerator.

Plum-Lavender Crisp

Plums pair beautifully with lavender. When used judiciously, it creates an enhanced flavor that is slightly floral. Your guests may try to guess what the other flavor is in the crisp without thinking they're eating their grandmothers' soap.

Makes 6 to 8 servings

For the streusel topping:
¾ cup (102 g) whole wheat flour

½ cup (1 stick; 113 g) unsalted butter, cold

½ cup (109 g) brown sugar

½ cup (50 g) rolled oats

For the fruit filling:
10 to 12 medium (about 2 pounds or 908 g) plums

¼ cup (34 g) all-purpose flour

¼ cup (50 g) granulated sugar

¼ cup (55 g) brown sugar

¼ teaspoon kosher salt

½ vanilla bean, split and scraped, or ¼ teaspoon ground vanilla bean, or ½ teaspoon vanilla paste

2 teaspoons ground lavender buds

Vanilla ice cream or whipped cream (optional)

1. Preheat the oven to 350 degrees F (175 degrees C).

2. *To make the streusel topping,* combine the flour, butter, sugar, and oats in a medium bowl. With your hands, work the butter into the dry ingredients until the mixture starts to clump. The butter should be completely worked in and the mixture should be moist. Set aside.

3. *To make the fruit filling,* cut the plums into ¼-inch-thick slices. You should have about 5 cups of sliced plums. Combine the plums and the remaining filling ingredients in a large bowl and toss to coat. Transfer the fruit mixture to a 2-quart baking dish.

4. Spread the streusel topping evenly over the fruit. Bake the crisp at 350 degrees F (175 degrees C) for 40 to 45 minutes, or until the fruit bubbles and the topping is browned. (If the topping browns too quickly, place a large piece of foil loosely on top to shield it from direct heat. Do not tuck in the foil.)

5. Serve the crisp warm or at room temperature with vanilla ice cream or whipped cream.

Fromage Blanc Cheesecake with Brown-Butter-Braised Plums

This unorthodox cheesecake marries the tangy flavor of *fromage blanc* to the deep flavor of the brown-butter-braised plums. Three simple techniques are brought together here in this fall dessert: (1) the key to cheesecake is not to overmix the batter so that the cake stays dense and puffs minimally; (2) brown butter acts as a base for the braising liquid, introducing deep, rich flavor; and (3) the braising technique, where the plums are partially submerged in the liquid, can be used for all stone fruits such as peaches, nectarines, apricots, and cherries.

Makes 6 servings

For the cheesecake:
1 cup (227 g) goat fromage blanc or chèvre, room temperature (see note, page 130)

⅓ cup (76 g) cream cheese, room temperature

¼ cup (28 g) mascarpone, room temperature

½ vanilla bean, split and scraped, or ¼ teaspoon ground vanilla bean, or ½ teaspoon vanilla paste

½ cup (100 g) sugar

2 eggs, room temperature

For the braised plums:
6 medium plums or pluots

¼ cup (57 g) unsalted butter

½ cup (109 g) light brown sugar

2 allspice berries

1 cup (227 mL) white wine, such as Riesling or viognier, plus more as needed

1 cup (227 mL) apple or pear cider, plus more as needed

1. *To make the cheesecake,* preheat the oven to 325 degrees F (165 degrees C). Grease a six-cup muffin tin. Have ready a 2-inch-deep pan large enough for the muffin tin to sit in easily. Boil enough water to fill the deep pan halfway to create a hot-water bath in which to bake the cheesecakes. Keep the water hot while you are mixing the cheesecake batter.

2. Combine the fromage blanc, cream cheese, mascarpone, and vanilla bean pulp in the bowl of a stand mixer

fitted with the paddle attachment. Mix on low speed until well combined, scraping two to three times. Be sure the mixture is smooth before moving to the next step.

3. Add the sugar and mix until combined, scraping once. Add the eggs one at a time, mixing and scraping well until the mixture is smooth.

4. Portion the cheesecake batter between the muffin cups. Pour some of the boiling water into the water-bath pan, and set the muffin tin in the water. Continue adding boiling water (be careful not to spill it into the cheesecakes) until the water reaches halfway up the muffin tin.

5. Place the water bath with the cheesecakes into the oven, and bake the cheesecakes in the water bath for 25 minutes or until just set. Tap the side of the muffin tin to confirm the cheesecakes still wiggle just a bit at the center of the cakes. Remove the water-bath pan from the oven, and set the muffin tin on a rack to cool for 10 minutes.

6. Invert the cheesecakes onto a parchment-lined baking sheet. (They should slip easily from the pan.) Refrigerate the cheesecakes, covered, until cold. They can be made 1 day ahead.

7. *To make the braised plums,* cut the plums in half from stem end to bottom and remove the pits.

8. Brown the butter over medium-high heat in a 10-inch skillet until it is fragrant and brown bits are evident on the bottom of the pan. Be sure to take this step far enough to get the deep brown-butter flavor, but not so far that it burns.

9. Add the plums cut side down, and brown them quickly in the butter over medium-high heat, about 1 minute. Turn the plums and cook another minute. Remove the plums to a plate. →

10. Add the brown sugar and allspice to the pan and stir to dissolve the sugar, about 1 minute. Add the white wine and cider. Bring the braising liquid to a simmer over medium-high heat, and add the plums, cut side down. The plums should be half-submerged in the braising liquid. Add a little more wine, cider, or water to increase the liquid if necessary. Bring the liquid back to a simmer, and adjust the temperature so the liquid is simmering gently. Cover the pan and braise the plums for 10 to 20 minutes, depending on the ripeness of the plums. The skins of the plums will loosen and shrivel some, but the fruits look best if kept mostly intact.

11. Serve two plum halves with one cheesecake and a little of the braising liquid. Hold extra plums refrigerated in the liquid for up to 5 days. (The braising liquid will need to be warmed slightly before using after refrigeration because the butter in the liquid will harden.)

Note: Goat fromage blanc and chèvre, both made from goat milk, are nearly identical in flavor. However, because chèvre is drained for a longer period, fromage blanc contains more moisture. If you use chèvre, I recommend adding 1 tablespoon of whole milk or cream to the cheesecake mixture.

APPLES

For many years, my mother made her apple pies with Transparent apples, an heirloom apple we picked in late August from my grandmother's sprawling tree in south Seattle. Boxes and boxes of apples were loaded into our 1970s station wagon and transported back to our house, where we spent days washing, peeling, coring, slicing, and freezing. We made and canned apple sauce and packed bags of sliced apples into the freezer for the coming year. In the autumn and deep into the winter, there were many, many apple pies.

Apples are one of the most interesting fruit varieties, and their history is long. (I recommend reading the first chapter of Michael Pollan's *The Omnivore's Dilemma*, which provides a fascinating history of apples in the United States.) Fortunately, growers of heirloom apples have revived many varieties that used to grow in the United States and Europe, dating back to the sixteenth century, thus giving us more diverse flavors and textures. A few favorites are Calville Blanc, Winesap, Northern Spy, King David, Roxbury Russet, and Arkansas Black. Many great-tasting apples have been "born" in the last fifty years, but, for me, the heirloom varieties possess unique flavors I don't find in the newer apples. The best way to learn about your local heirloom apple varieties is to talk with an orchardist or knowledgeable market staff. They can tell you which apples are best for baking, saucing, freezing, eating fresh, and storing.

Rosemary Apples in Crepes with Rum Caramel Sauce

This dessert is a favorite of mine; the combination of the apples, caramel, and rosemary folded into crepes can't be beat. Don't pass up this dessert because of its extra components. The crepes and caramel can be made ahead. Then the dessert is just a simple slice and sauté of the apples. Like many other recipes in this book, just the apple preparation is a dessert on its own. As an autumn or winter dessert, this one is indispensable.

Makes 6 servings

½ recipe Crepes (page 72)

1 recipe Rum Caramel Sauce (recipe follows)

4 medium apples (tart-sweet, such as Jonagold or Pink Lady)

3 tablespoons unsalted butter

2 tablespoons light brown sugar or honey

½ vanilla bean, split and scraped, or ¼ teaspoon ground vanilla bean, or ½ teaspoon vanilla paste

1 teaspoon fresh chopped rosemary

1. Prepare the crepes, then the caramel sauce.

2. Cut the apples into ⅛-inch-thick slices (do not peel them). Gather all the other ingredients, and set them near the stove before you start cooking. This is a quick cook!

3. Melt the butter over medium-high heat in a large skillet until it is fragrant and brown bits are evident on the bottom of the pan. Be sure to take this step far enough to get the deep brown-butter flavor, but not so far that it burns.

4. Add the sugar and vanilla bean, and cook until the sugar dissolves, about 2 minutes. Increase the heat to high, and add the apple slices. Sauté the fruit quickly, tossing or stirring occasionally. Add the rosemary and toss with the apples to combine. Remove the apples from the heat.

5. Place one crepe on each dessert plate. Divide the apples between the six crepes. Fold the crepes in half and then half again. Drizzle a few tablespoons of rum caramel over the crepes. Serve immediately. →

RUM CARAMEL SAUCE

Makes 8 to 10 servings

⅔ cup (133 g) sugar
⅓ cup (75 g) honey
⅓ cup (76 mL) water
1 cup (227 mL) heavy cream
¼ cup (57 mL) dark rum (such as Myers's Original Dark Rum)
2 tablespoons unsalted butter
¼ teaspoon kosher salt

- Combine the sugar, honey, and water in a small pot. Stir gently to combine. Wash down any sugar on the sides of the pot with a wet pastry brush. Bring the sugar mixture to a boil over high heat. (Do not stir the sugar after brushing down the sides, as stirring causes the sugar to crystallize. Once the sugar starts to caramelize, you can swirl the pot gently to even out the color.) Continue to cook until the mixture is a dark caramel color, about 300 degrees F (150 degrees C) if you are using a thermometer. I do this step by sight, watching and waiting for a deep amber color and caramel aroma. Often you will see a puff of smoke appear above the pot, which is the exact time to pour in the cream for a deep caramel.

- When the sugar mixture reaches the caramel stage, remove the caramel from the burner and carefully whisk in the cream a little at a time. (The cream generates a lot of steam.) Whisk in the rum, butter, and salt. If the caramel still has some lumps, return the pot to the burner and whisk the caramel over the heat until it is smooth. Pour the caramel into a metal bowl to cool.

- This recipe makes more than you need for the crepes, but I don't expect to hear complaints of how to use the leftovers. It would be a great accompaniment to the Brandied Apple Pie (page 138) or the Apple Upside-Down Cake (page 142). Store the caramel in the refrigerator for up to 2 weeks. Warm it gently before serving.

Brandied Apple Pie with Ginger Streusel

For diverse flavor and texture, I always use two to four apple varieties when making an apple pie. A dependable combination of tart, tart-sweet, and sweet apples is Granny Smith, Fuji or Jonagold (or other tart-sweet), and Golden Delicious. The tart apple holds its shape while cooking, while the other two break down a bit more in the pie. If you frequent the farmers' markets, ask around for the farmers' favorite pie apples. This pie recipe has become one of my signature desserts. It is softly spiced, and the addition of brandy lends a pleasing depth to the pie without overpowering the apples.

Makes 6 servings

½ recipe Flaky Pie Dough (page 12)

For the streusel:

½ cup (65 g) all-purpose flour

¼ cup (53 g) brown sugar

¼ cup (25 g) rolled oats

2 tablespoons finely chopped candied ginger

¼ cup (57 g) unsalted butter

For the filling:

7 medium apples (a combination of tart, tart-sweet, and sweet)

¼ cup (57 g) brandy

1 tablespoon apple cider vinegar

½ cup (100 g) sugar

½ cup (71 g) dried currants

¼ cup (33 g) all-purpose flour

½ teaspoon ground ginger

¼ teaspoon ground cinnamon

¼ teaspoon kosher salt

1½ tablespoons unsalted butter

Vanilla ice cream or whipped cream (optional)

1. Prepare the pie dough.

2. *To make the streusel,* combine the flour, brown sugar, oats, and candied ginger in a medium bowl. Cut the butter into ½-inch pieces, and add it to the other ingredients. With your hands or a pastry blender, work the butter into the dry mixture until the streusel starts to hold together and no pieces of butter are evident. The streusel should be neither dry nor sticky, but will hold together with a gentle squeeze. Hold the streusel at room temperature while you assemble the pie.

3. *To make the filling,* peel and slice the apples ⅛ inch thick. Toss together the sliced apples, brandy, and vinegar in a large bowl. Add the sugar, currants, flour, ginger, cinnamon, and salt and toss to combine.

4. *To assemble the pie,* roll out the bottom crust, line the pie pan, and crimp the edges. (See How to Roll Pastry [Pie] Dough, page 13.) Pile the apple filling into the bottom crust, mounding 1 to 2 inches above the edge of the pie pan. Dot the top of the fruit with the butter.

5. Gently distribute the streusel topping over the apple filling, covering the apples. I like to leave the streusel fairly loose on the top of the pie, so it has a nice textured appearance when baked. Chill the finished pie in the freezer for 30 minutes before baking. Meanwhile, preheat the oven to 400 degrees F (200 degrees C).

6. Bake the pie for 20 minutes, or until the edge of the crust is dry and starting to color. Reduce the temperature to 350 degrees F (175 degrees C) and continue baking for 30 to 40 minutes more, or until the pie is brown and the juices are bubbling. (Crumb-topped pies can start to brown too quickly while they are baking. If this happens, lay a piece of foil loosely over the top of the pie. Do not tuck it around the pie. The foil acts as a shield and protects the top from getting too dark, but still allows the pie to continue baking.)

7. Cool the pie to room temperature or slightly warm before serving with vanilla ice cream or whipped cream.

Note: Whenever I make streusel, I make plenty. I freeze the extra in plastic bags so that it's available for a quick crisp anytime throughout the season.

^ Brandied Apple Pie with Ginger Streusel > Apple Upside-Down Cake with Spelt and Rye Flours

Apple Upside-Down Cake with Spelt and Rye Flours

The spelt and rye flours, as well as a variety of sweeteners, deepen the flavor of this simple upside-down cake. There is more to whole grain flours than just wheat. Spelt (an ancient wheat) acts like a soft whole wheat pastry flour, and rye flour's low gluten content, as well as its earthy flavor, adds both interesting texture and flavor. I use several types of sweeteners to achieve a deep, round flavor without it being too sweet. (For more information on sweeteners, see Sugar, page x.) Any firm cooking apple will work here, but I prefer the sweet and dependable Golden Delicious because its simple flavor melds well with the earthy taste of this cake.

Makes 8 to 10 servings

For the apple topping:
4 to 5 medium apples

¼ cup (55 g) brown sugar

¼ cup (57 mL) honey

4 tablespoons unsalted butter

2 tablespoons maple syrup

⅛ teaspoon kosher salt

For the cake:
½ cup (1 stick; 113 g) unsalted butter, at room temperature

½ cup (100 g) plus 1 tablespoon granulated sugar, divided

2 tablespoons brown sugar

2 tablespoons honey

¼ teaspoon kosher salt

3 eggs, separated, at room temperature

1 teaspoon vanilla extract

1½ cups (180 g) whole spelt flour

¾ cup (83 g) medium or light rye flour

2 teaspoons baking powder

¾ teaspoon ground cardamom

½ cup (113 mL) whole milk

1. Preheat the oven to 350 degrees F (175 degrees C).

2. *To make the apple topping,* peel and cut the apples to ½-inch-thick slices. Combine the brown sugar, honey, butter, maple syrup, and salt in a 10-by-2-inch round cake pan. Cut the butter into eight pieces and add to the pan. Place the pan in the oven to melt the butter. Once the butter is melted, remove the pan from the oven and stir to combine the ingredients. Add the apples, and arrange them so that the slices are in one fairly snug spiraled layer on the bottom of the pan. Set aside.

3. *To make the cake,* cream the butter, ½ cup of the granulated sugar, brown sugar, honey, and salt in a stand mixer fitted with the paddle attachment until light and fluffy, about 5 minutes, scraping once. Add the egg yolks one at a time, scraping well after each addition. Mix in the vanilla.

4. Sift together the flours, baking powder, and cardamom. With the mixer on low speed, add half the milk to the butter mixture, then follow immediately with half the flour and mix until almost combined. Scrape well. Add the rest of the milk and the remaining flour, scraping and mixing again until just combined. Move the batter to a large bowl so you can use the stand mixer bowl for the next step. Clean the mixer bowl well before whipping the egg whites.

5. Whip the egg whites with the remaining 1 tablespoon of sugar to medium-stiff peaks. With a spatula, gently fold one-third of the egg whites into the cake batter until almost combined. Fold in the remaining egg whites until fully combined.

6. Drop the batter in five mounds on top of the apples, and gently spread the cake batter in the pan, using light strokes to level the batter. Bake the cake for 45 to 55 minutes, or until it springs back when touched. (You may see some apple juices bubbling up on the sides.)

7. Cool the cake for 10 to 15 minutes. Place a serving plate on top of the cake so that you can invert the cake onto the serving plate. Gently remove the cake pan. Cool the cake until just warm or it reaches room temperature. Serve with slightly sweetened whipped cream.

Cider-Braised Apples with Warm Winter Spices

I remember my mother in the kitchen cooking for all of us on cold fall nights. We always had dessert—most often some of her canned fruit or a simply prepared fruit, like these baked apples. Filled with dried currants, spices, and a little sugar, and surrounded with apple cider and juices from the apples themselves, this is a quintessentially autumn dessert and a warm, comforting way to end a meal. My mother often used Rome apples because their red-streaked flesh cooked down soft and flavorful, and their deep red and round shape was beautiful, even after cooking. If I can't find Romes, I reach for Jonagold, Liberty, Arkansas Black, or any mostly red sweet or tart-sweet apple.

Makes 6 servings

6 medium sweet or tart-sweet apples, cored and peel intact

½ cup (71 g) dried currants

½ cup (about 70 g) chopped dried figs, apricots, or prunes (or a combination of the three)

6 tablespoons light brown sugar

½ teaspoon ground cinnamon

½ teaspoon ground ginger

¼ teaspoon kosher salt

⅛ teaspoon ground nutmeg

2 cups (454 mL) apple cider

1. Preheat the oven to 350 degrees F (175 degrees C). Set the apples stem end up in a 9-by-13-inch pan.
2. Combine the currants, dried fruit, sugar, cinnamon, ginger, salt, and nutmeg in a medium bowl and mix well.
3. Heat the apple cider in a medium pot to boiling, then pour the cider over the apples. (The cider helps keep the apples moist while they are baking and provides even more flavor.)

4. Spoon the dried fruit filling into the center of each apple, packing it gently and filling to slightly above the top.

5. Cover the apples with foil, leaving a small air gap above them. Bake the apples for 45 minutes to 1 hour, or until they are tender but not falling apart.

6. Serve the apples warm with some of the cider spooned over the tops.

PEARS

Pears are a humble fruit, falling in line behind apples, the stars of autumn. I'm not sure where my intense love of pears came from (although we ate a lot of pears growing up—fresh and home-canned), but give me a choice of any fruit to eat fresh, out of hand, and I will choose a perfectly ripe yellow Bartlett pear.

However, Bartlett pears are not always my choice for cooking and baking. I find them perfect for pies, crisps, compote, and sauces because they tend to break down easily and they have a pleasant tart-sweet flavor. For poaching, roasting, and braising pears, look for Bosc, Anjou, Comice, Forelle, Starkrimson, or the petite Seckel pear. I go to the Bosc for roasting and poaching (partially because of its handsome, slender shape), Anjou and Comice for most cooking techniques, and the little Seckel for the most beautifully cooked mini pears. Pears don't get enough time in the spotlight, so here's to pears catching up with apples as a prime autumn fruit.

^ Pear and Fig Pie > Burgundy-Poached Pears with Mulled-Wine Syrup

them under all the way around the pie, pressing the dough onto the rim of the pie pan, or just inside the pie pan if your pan doesn't have a flat edge. Now crimp the edge with a fork or between your fingers to create a seal and a pretty edge.

6. Chill the finished pie in the freezer for 30 minutes before baking. This step will keep the dough from melting before it bakes and help achieve a baked crust on the bottom. While the pie is in the freezer, preheat the oven to 400 degrees F (200 degrees C).

7. Before baking, brush the top of the pie with the milk and sprinkle sugar over the top. Bake the pie for 20 minutes, or until the top of the pie is dry and just starting to color. Reduce the oven temperature to 350 degrees F (175 degrees C) and continue baking for 30 to 40 minutes more, or until the pie is brown and bubbling. The pie should be golden on top, and you should be able to hear or see the juices bubbling in the middle. (If you're unsure whether the fruit is cooked, insert a dinner knife into one of the slits. No resistance means the fruit is soft and cooked.)

8. Cool the pie to room temperature and serve with vanilla ice cream or whipped cream.

^ Pear and Fig Pie > Burgundy-Poached Pears with Mulled-Wine Syrup

PEARS

Pears are a humble fruit, falling in line behind apples, the stars of autumn. I'm not sure where my intense love of pears came from (although we ate a lot of pears growing up—fresh and home-canned), but give me a choice of any fruit to eat fresh, out of hand, and I will choose a perfectly ripe yellow Bartlett pear.

However, Bartlett pears are not always my choice for cooking and baking. I find them perfect for pies, crisps, compote, and sauces because they tend to break down easily and they have a pleasant tart-sweet flavor. For poaching, roasting, and braising pears, look for Bosc, Anjou, Comice, Forelle, Starkrimson, or the petite Seckel pear. I go to the Bosc for roasting and poaching (partially because of its handsome, slender shape), Anjou and Comice for most cooking techniques, and the little Seckel for the most beautifully cooked mini pears. Pears don't get enough time in the spotlight, so here's to pears catching up with apples as a prime autumn fruit.

Pear and Fig Pie

This pie almost didn't make the cut for this book. But as I was finishing the first draft, a discerning teaching colleague told me how she and her husband had recently made this pie and *loved* it! She went on about the flavors, how they melded. "Mmm," she said. "That Pear and Fig Pie!" And she smiled. I took it as a sign that it should go back in the book.

Makes 6 servings

1 recipe Flaky Pie Dough (page 12)

1 cup (140 g) dried Mission figs, stemmed and quartered

6 medium pears (such as red or yellow Bartlett, Anjou, or Comice, or a combination)

2 tablespoons apple cider vinegar

½ cup (100 g) sugar, plus extra for sprinkling

¼ cup (34 g) all-purpose flour

½ teaspoon ground ginger

¼ teaspoon ground cardamom

¼ teaspoon kosher salt

2 pinches ground cloves

3 tablespoons whole milk, for finishing

Vanilla ice cream or whipped cream (optional)

1. Prepare the pie dough.

2. To prepare the figs, place them in a colander and pour boiling water over them. Shake off the excess water.

3. Peel and cut the pears into ⅛-inch-thick slices. Toss together the pears, figs and vinegar in a large bowl. Add the sugar, flour, ginger, cardamom, salt, and cloves and toss to combine.

4. Roll out the pie dough. (See How to Roll Pastry [Pie] Dough, page 13.) Roll out the bottom crust to about ⅛ inch thick and line a 9-inch pie pan. Trim the edge of the bottom crust with a scissors or a knife to ¼ inch over the edge of the pie pan. Toss the pear filling a few more times, then spoon it into the pie pan, mounding the filling 1 to 2 inches above the edge of the pie pan.

5. Roll out the top crust to about ⅛ inch thick. Place the crust on top of the pie, and trim the edge flush with the bottom crust. Pick up the overhanging edges, and turn

Burgundy-Poached Pears with Mulled-Wine Syrup

This is an elegant dessert. I employ poaching here again, but this time with deep red wine and spices. You'll think the holidays are just around the corner as the fragrance of the spiced poaching liquid fills your kitchen. Bosc pears are my favorite for this recipe because of their graceful shape, but Comice or any firm pear will work. The pears are best poached one day ahead; as they sit, they absorb more flavor and color, turning a deep pink to red. A stunning effect.

Makes 4 servings

For the poached pears:

4 cups (about 1 L) Burgundy wine or other French red wine

4 cups (about 1 L) water

1 cup (227 mL) port (or replace with the same red wine, if you don't have port)

2 vanilla beans

2 cinnamon sticks

2 strips orange zest

5 medium, firm pears

1. *To make the poached pears,* combine the wine, water, port, vanilla beans, cinnamon sticks, and orange zest in a large pot. Bring to a boil over high heat, and then reduce the heat to low and let the poaching liquid steep for at least 30 minutes. (If you have an hour, that's even better.) With an apple corer, gently core the interior of each pear from the bottom, inserting the corer just far enough toward the stem to remove the core. Remove the corer (the core may not come out with it), and use a paring knife to gently work out the interior core. (If you don't want to attempt this, simply peel and halve the pears and then carefully remove the stem and core of each of the pear halves with a small melon baller or the tip of a spoon.) Once you have removed each pear core, gently peel the pear, leaving the stem intact.

2. When the poaching liquid is ready, add the pears. Cut a piece of round parchment and lay it on top of the pears. Use a cake pan to keep the pears submerged in the liquid. Poach the pears until they are knife-tender but not falling apart. (I like to gently squeeze them to check the give. The pears should be slightly firm.) For the best color, poach the pears the day before you plan to serve them. Hold the pears in the poaching liquid refrigerated until you are ready to use them—up to 1 week. This recipe includes one extra pear because I usually poach extra fruit in case one falls apart. In this case, you'll only need four of the five pears. The extra pear would make the perfect pairing to salad greens.

3. *To make the mulled-wine syrup*, pour 1½ cups (340 mL) of the poaching liquid into a medium pot and bring to a boil over high heat. Lower the heat to a simmer, and reduce the poaching liquid to a light syrup, 20 to 30 minutes or until the liquid is reduced by two-thirds. Cool the syrup before serving. The remaining poaching liquid can be used twice more to poach additional fruit. It will hold in the refrigerator up to 2 weeks.

4. When you are ready to serve, place four pears on a paper towel, stems straight up, to catch any extra juices, then move the pears to serving plates. Drizzle a few tablespoons of the syrup onto each plate and serve.

Maple and Pear Panna Cotta

Panna cotta (Italian for "cooked cream") is a traditional dessert and might be one of the simplest to make. The mistake many cooks make is adding too much gelatin. Serving panna cotta in a vessel means we can keep the gelatin content low and achieve a creamier texture. The pear compote pairs well with the deep maple sugar, giving it almost a buttery flavor. I prefer Bartlett pears here, but any pear will work. Maple sugar is available in some grocery stores and online, but you can replace it here with real maple syrup.

Makes 6 servings

For the panna cotta:
1 tablespoon water

1 packet (about 2½ teaspoons) powdered gelatin

2 cups (454 mL) heavy cream

½ cup (113 mL) whole milk

¼ cup (50 g) maple sugar, or ¼ cup (57 mL) real maple syrup

2 tablespoons granulated sugar

For the compote:
3 to 4 medium pears

3 tablespoons sugar

3 tablespoons water

Pinch ground cinnamon

Pinch ground nutmeg

Pinch ground cardamom

1. *To make the panna cotta,* combine the water and gelatin powder in a small bowl and let stand for 5 minutes to bloom.

2. Combine the heavy cream, milk, and sugars in a medium pot. Heat over medium heat until hot and the sugar is dissolved. Do not allow the mixture to boil.

3. Whisk about ½ cup (113 g) of the hot mixture into the bloomed gelatin until the gelatin is dissolved. Then whisk the gelatin mixture from the bowl into the pot. Strain the mixture through a fine-mesh sieve or cheesecloth. Pour the panna cotta into six small, deep bowls, such as 4-ounce ramekins, or use any small decorative bowls. Chill the panna cotta in the refrigerator for 2 hours

or until set. (If you are holding the panna cotta overnight, after they set, carefully stretch plastic wrap over them.)

4. *To make the pear compote,* peel and cut the pears to a ½-inch dice. You should have about 2 cups (454 g) of pear pieces. Combine the pears with the remaining ingredients in a medium pot. Bring to a gentle simmer over medium heat. Reduce the heat to low and continue to simmer gently until the pear pieces soften but are not falling apart, 5 to 8 minutes. Transfer to a bowl and cool the pear filling in the refrigerator.

5. When you are ready to serve the panna cotta, spoon 2 tablespoons of the compote over one side of each bowl of panna cotta. Serve immediately.

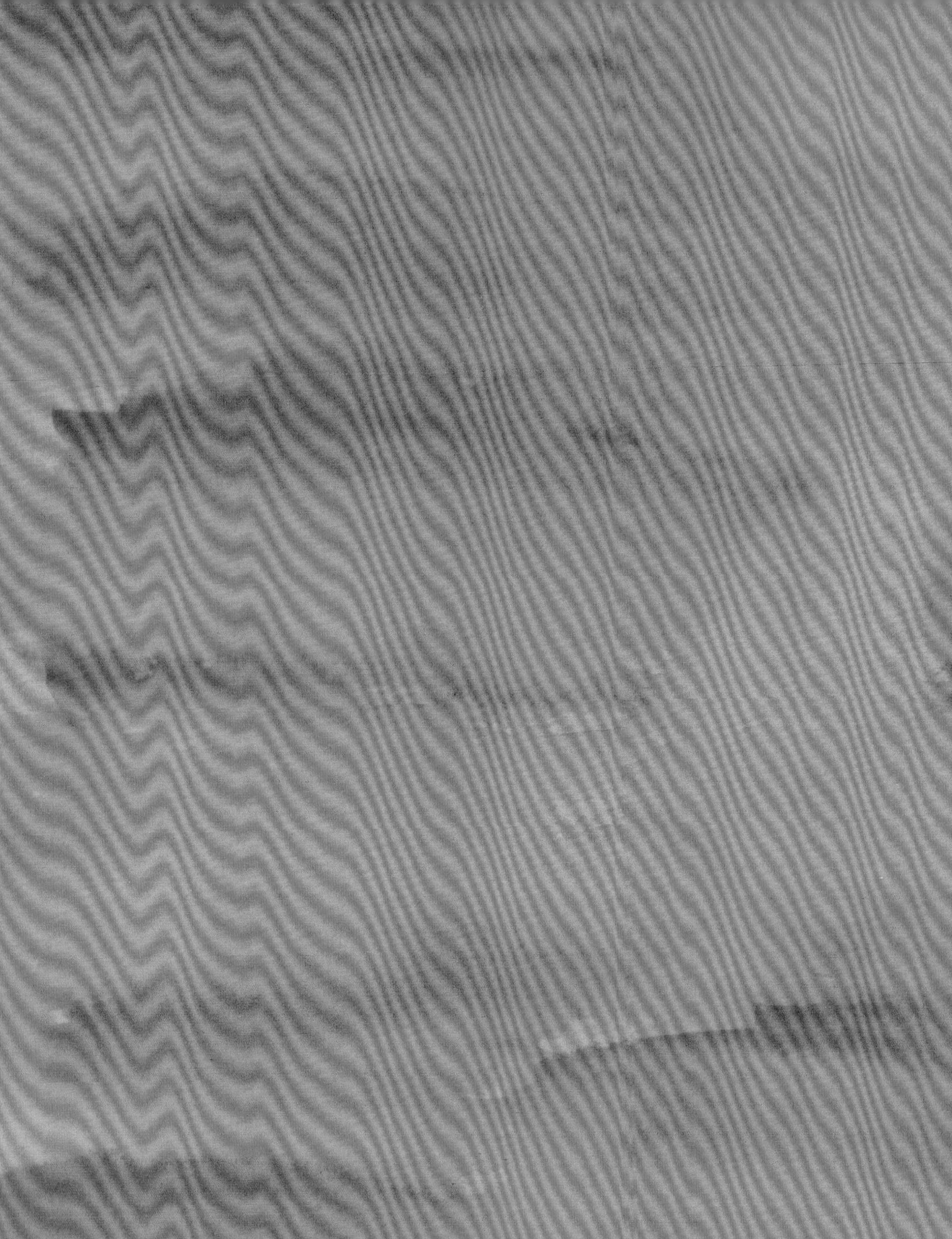

WINTER

CRANBERRIES 159

Mom's Cranberry-Walnut Pie 160

Apple-Cranberry Tarte Tatin 164

Cranberry and Bay Upside-Down Cake 166

CITRUS 169

Meyer Lemon Buttermilk Sherbet 170

Almond Cake with Warm Citrus and Thyme 173

Orange and Rye Madeleines 175

Orange, Rosemary, and Hazelnut Brittle 177

DRIED FRUITS 181

Grandma's Dark Holiday Fruitcake 182

Apricot-Almond Scones 186

CRANBERRIES

We ate a lot of cranberry desserts, salads, and sauces when I was growing up. I've never had canned cranberry sauce; Mom always made it from scratch. She also made a wonderful gelatin salad with ground fresh cranberries and walnuts, which I still crave. And of course, cranberry bread was a Thanksgiving staple. Like me, she always stored a few bags in the freezer for a late-winter cranberry treat.

The wild Washington coast is home to over two thousand acres of cranberries, but that's just a fraction of the more than forty thousand acres in the United States. Cranberries have a long history in the United States as a native fruit used for both food and dyes. It's a sign of the beginning of the holiday season when I see them in the market. As soon as the price is reasonable, I snatch up several pounds to store in my freezer because they are hard to come by at other times of the year. I love to cook them with as little sugar as I can stand and mix the compote (essentially sauce) into my yogurt or just eat it plain. (Cranberries are a great source of vitamin C!) It's simple to drop in a vanilla bean, a cinnamon stick, or a slice of fresh ginger to give this super-tart berry a roundness of flavor.

Mom's Cranberry-Walnut Pie

Because my mom and I are huge fans of tart fruits like cranberries, this pie is a favorite in our family. My mom often made it for Christmas dinner, and although I don't know the origin of the recipe, it has lived in her recipe clippings for forty-some years. She and I prefer the walnuts, but you could substitute pecans. Be sure to finish the pie with the glaze to give it a glistening look, and a little pie decor on top is not out of place here.

Makes 6 servings

½ recipe Flaky Pie Dough (page 12)

3 cups (340 g) fresh cranberries

2 medium sweet apples, peeled and coarsely grated

1 cup (120 g) walnuts

⅔ cup (133 g) sugar

2 tablespoons all-purpose flour

1 tablespoon unsalted butter

3 tablespoons apricot jam

2 tablespoons water

1. Prepare the pie dough and roll out the crust. (See How to Roll Pastry [Pie] Dough, page 13.) Form the pie crust in the pie pan, crimping the edges, and freeze it until you are ready to fill it (for at least 30 minutes until firm). Meanwhile, preheat the oven to 375 degrees F (190 degrees C).

2. Combine the cranberries, grated apples, and walnuts in a food processor and pulse two to three times until roughly chopped. Transfer the mixture to a large bowl, and add the sugar and flour, then gently toss to combine. Break up the butter into eight pieces, and toss them together with the pie filling.

3. Remove the pie crust from the freezer, and add the filling. Level the filling gently, but don't press it down.

4. Bake the pie for about 15 minutes, or until the edge of the crust is dry and set. Lower the temperature to 350 degrees F (175 degrees C), and continue baking until the filling is tender and bubbling and just a few spots of brown are evident on the top, 25 to 30 minutes. (If the pie is getting too dark, lay a loose piece of foil over the top to shield it.) Cool the pie on a rack.

5. Combine the apricot jam with the water in a small pot. Bring to a simmer. Strain the glaze through a fine-mesh sieve to remove the apricot pieces. Brush the glaze onto the cooled pie, being careful not to glaze the crust.

6. Serve the pie at room temperature. This pie keeps well at room temperature for 2 days or covered in the refrigerator for 5 days.

< Mom's Cranberry-Walnut Pie ^ Apple-Cranberry Tarte Tatin

Apple-Cranberry Tarte Tatin

This classic tarte tatin features cranberries, which, surprisingly, hold up during the cooking process. I love both the visual result and the contrast of the tart cranberries next to the caramelized, sweet apples. I prefer Golden Delicious apples here, but Opal, Fuji, or Jonagold also work well. This classic dessert is best eaten slightly warm and doesn't suffer by adding a drizzle of heavy cream or whipped cream.

Makes 6 to 8 servings

½ recipe Flaky Pie Dough (page 12), or store-bought puff pastry

6 tablespoons cane sugar, such as turbinado, brown, or demerara

¼ cup (57 mL) water

1 tablespoon cider vinegar

Pinch kosher salt

¼ cup (57 mL) apple or pear brandy, such as Calvados

4 tablespoons unsalted butter

6 medium, firm, sweet apples

1½ to 2 cups (150 to 200 g) fresh or previously frozen cranberries

Heavy cream or whipped cream (optional)

1. Prepare the pie dough.

2. In a heavy 10-inch skillet (preferably a light-colored skillet so you can see the caramel easily), stir together the sugar, water, vinegar, and salt and bring to a boil over high heat. Continue to cook without stirring until the sugar caramelizes and turns a dark amber caramel. Add the brandy and butter to the pan and whisk to combine, bringing the mixture back to a simmer. Remove the pan from the heat. Let the caramel set in the pan off the heat for 5 minutes. (It's OK if the caramel hardens.)

3. Peel, core, and quarter the apples. Place the apple quarters in the caramel with the peeled side of the apple pieces facing the bottom of the pan and the interior of the apple facing up. Fit them closely together in concentric circles, but don't overlap them. Sprinkle the cranberries over the sliced apples, pushing them down

between the apples so that they show when the tart is inverted. (Fit in as many cranberries as possible, up to 2 cups, or 200 g.) Finally, thinly slice any remaining apple quarters and sprinkle them on top.

4. Return the pan to medium-high heat, cover, and simmer the caramel and juices (reduce the heat, but keep the liquid at a simmer) until the apples start to soften, about 20 minutes.

5. While the apples are cooking, preheat the oven to 400 degrees F (200 degrees C).

6. Roll out the pie dough to a 10½-inch round. (See How to Roll Pastry [Pie] Dough, page 13.) Trim the edge of the pastry dough with a knife. Place the dough on a parchment-lined baking sheet in the fridge until you are ready to use it.

7. Remove the lid from the apples, and let most of the liquid boil off so that just the caramel remains on the bottom of the pan. (Tip the pan to check that most of the liquid has boiled off.) Remove the pan from the heat.

8. Lay the dough over the apples in the pan, curving the edges of the dough slightly down. Bake the tart for 45 to 50 minutes, or until the crust is firm and dark golden brown. (The crust is likely to be soggy if it is underbaked.)

9. Let the tart cool to warm, about 10 minutes. Carefully invert the tart onto a large plate. (If the tart has cooled too much to remove it from the pan, put the pan back on medium-high heat to warm the bottom and loosen the apples.) Serve with a drizzle of heavy cream or whipped cream.

Cranberry and Bay Upside-Down Cake

Living in a cranberry state, I want to use this fruit all the time. With their deep red color and tart flavor, these easy-to-use berries make striking desserts. I was first inspired by a cranberry upside-down cake from Alice Waters's book *Fruit*. Upside-down cakes always need a fruit that can stand the heat at the bottom of the pan, and cranberries carry it off. The bay leaf is an added flavor that shouldn't overwhelm, but just add an interesting touch of earthiness and mystery.

Makes 8 servings

½ recipe Classic White Cake batter (page 94)

6 tablespoons light brown sugar

6 tablespoons granulated sugar

6 tablespoons unsalted butter

2 bay leaves

⅛ teaspoon kosher salt

3½ cups (450 g) fresh cranberries (or thawed if previously frozen)

Whipped cream (optional)

1. Prepare the cake batter, but do not bake it. Set the batter aside in the mixing bowl.

2. Preheat the oven to 350 degrees F (175 degrees C).

3. In a 9-inch round cake pan, combine the sugars, butter, bay leaves, and salt. Put the pan on the stove over low heat to melt the butter. (You could also do this step in the oven. I recommend using the oven if you are using a nonstick cake pan.) Increase the heat to medium, and cook the sugars until they bubble. Remove the pan from the heat, and steep the bay leaves in the caramel for 30 minutes, stirring once.

4. Remove the bay leaves from the caramel (you should be able to smell the bay leaf in the caramel), and add the cranberries to the pan. (Do not stir the cranberries

into the caramel.) Spoon the cake batter into the pan in four mounds, and gently spread it to the edges. Give the pan a few taps on the counter to settle the cake.

5. Bake the cake for 40 to 45 minutes, or until the middle of the cake springs back when touched. Let the cake stand for 20 minutes at room temperature before inverting it onto a large serving plate. If the cranberries are falling off the edge after inverting, wrap a strip of parchment around the edge (like a collar) to hold the cranberries while the cake cools.

6. Remove the parchment collar, and serve the cake with whipped cream.

CITRUS

It feels strange when citrus comes into the markets in the Pacific Northwest because it means the sun is shining somewhere, at least in that sunny place where citrus comes from, down south in California or Texas or Florida. But citrus also indicates that winter is right around the corner (or already here), and we're about to settle in for a cold, wet period that needs some bright spots. I'll indulge in sweet and tangy satsumas, tangerines, and Cara Cara oranges before Christmas. But the period after Christmas is my favorite citrus time of the year. I search for varieties that get overlooked, like pomelos or the intensely bitter Seville oranges (we love marmalade in our family). The popular Meyer lemons also arrive, and it's really the perfect time to enjoy grapefruit.

When picking citrus, I depend on the fragrance of the peel. And eating organic citrus can make a difference, especially when you are using the peels. Save the peels from a week of citrus, and make candied or dried citrus peel for enjoying later in the year.

Meyer Lemon Buttermilk Sherbet

Although standard lemons are available all year, make this sherbet when Meyer lemons are in season. The Meyer lemon—a somewhat less astringent lemon because it's a cross between a mandarin-pomelo and a citron—originated in China and was brought to the United States in the early twentieth century. Thankfully, Alice Waters made it popular again in the 1970s. The somewhat-odd combination of lemon and buttermilk in this sherbet may seem too intense on first consideration, but it works. I love the tanginess of this sherbet; it also makes a refreshing summer treat.

Makes 8 servings

2 cups (454 mL) buttermilk

1 cup (227 mL) corn syrup

½ cup (113 mL) freshly squeezed Meyer lemon juice (from about 6 Meyer lemons)

½ cup (100 g) sugar

⅓ cup (76 mL) whole milk

3 teaspoons lemon zest

1. Combine the buttermilk, corn syrup, lemon juice, sugar, and milk in a medium pot over medium heat. Cook, stirring frequently, until the sugar is dissolved. Stir in the lemon zest. Transfer the mixture to a metal bowl and refrigerate until cold, about 2 hours.

2. See How to Freeze Sorbet, Ice Cream, and Sherbet for freezing instructions (page 8).

Almond Cake with Warm Citrus and Thyme

This is another near and dear recipe of mine—one that I could enjoy every week. The marrying of various warm citrus (some sweet and some sour or bitter) with an herb is intoxicating. The flavors pair well with this almond cake, which is really a classic frangipane baked as a cake. Make the cake a day ahead for ease of putting it together later. The fruit sauté, like others in this book, is quick and simple. Just make sure to have all your ingredients ready to go before starting to cook.

Makes 6 servings

For the cake:
1 cup (142 g) almond paste, room temperature

1 tablespoon sugar

½ cup (1 stick; 113 g) unsalted butter, room temperature

3 eggs, room temperature

¼ cup (35 g) bread flour

For the citrus:
6 to 10 citrus fruits, such as navel oranges, grapefruit, clementines, and blood oranges (about 2 pounds)

3 tablespoons sugar or honey

2 teaspoons fresh thyme, roughly chopped

Pinch sea salt

2 tablespoons unsalted butter

1. *To make the cake,* preheat the oven to 350 degrees F (175 degrees C). Lightly grease the bottom of an 8-inch round cake pan, and line the pan with a circle of parchment that just fits the bottom.

2. Combine the almond paste and sugar in the bowl of a stand mixer. Using the paddle attachment, mix the almond paste and sugar on medium-low speed until the almond paste has broken into small pieces, about 1 minute. Add the butter, and cream the mixture on medium speed, scraping once, until light and fluffy and no pieces of the almond paste are visible, about 5 minutes. Add the eggs one at a time, scraping the bowl after each addition. Finally, add the bread flour and mix on low speed until just combined. Scrape as needed. →

3. Transfer the batter to the cake pan, smoothing the top. Bake the cake for 30 to 35 minutes until the cake is golden and springs back to the touch.

4. Cool the cake completely before removing it from the pan. For best flavor and texture, serve the cake at room temperature. You can also freeze the cake, triple wrapped in plastic wrap, for up to 1 month.

5. *To prepare the citrus* while the cake is baking, segment the large citrus (such as oranges and grapefruit) by cutting off the rind top and bottom of each fruit. Then cut off the rind on the sides from top to bottom, following the curve of the fruit. Using a paring knife, cut out each segment of the citrus from its membrane and put the pieces into a medium bowl, saving the juice as well. Small citrus (such as tangerines and clementines) can be peeled and the segments pulled apart. You should have about 3 cups (600 g) of segments. Refrigerate the citrus until you are ready to cook it.

6. Once the cake is baked and cooled, cut it into six wedges and place them on dessert plates. In a 10-inch sauté pan over medium-high heat, quickly warm the citrus and juice for 1 to 2 minutes. Add the sugar, thyme, and salt. Toss over the heat until combined and very hot, but don't allow the citrus to fall apart.

7. Remove the pan from the heat and add the butter. Toss gently to combine. Spoon the citrus over the slices of almond cake and serve immediately.

Orange and Rye Madeleines

Two of my favorite cookbook authors, Dorie Greenspan and David Lebovitz, inspired me on the madeleine road many years ago. Up until then, I didn't even have a madeleine pan. Madeleines are one of the simplest and most pleasing pastries you can make. A cross between a cookie and a cake, madeleine flavors and flours can be changed up for variety. Make the batter one day ahead to achieve the best trademark "hump," but even baking them the same day will give you a tender madeleine. Rye is my favorite whole grain flour to add to just about anything. The combination of orange and rye in this madeleine is the perfect winter accompaniment to a cup of hot tea.

Makes about 12 madeleines

½ cup (65 g) all-purpose flour

⅓ cup plus 2 tablespoons (50 g) light rye flour

½ teaspoon baking powder

½ cup (100 g) sugar

2 eggs, room temperature

Pinch kosher salt

½ cup (1 stick; 113 g) unsalted butter, room temperature, plus more for greasing

1 teaspoon orange zest

1 teaspoon vanilla extract, or the pulp of ½ vanilla bean

1. Sift the flours and baking powder in a small bowl. In a medium bowl, whisk together the sugar, eggs, and salt until frothy. Create a pomade with the butter by squeezing it through your fingers and into the bowl with the egg mixture, thus making the butter more pliable. Whisk the egg mixture until the butter pieces are small and evenly distributed. Stir in the orange zest and vanilla. Add the flour mixture and fold gently with a rubber spatula until just combined. (The batter will look a little lumpy because there will be small pieces of butter visible in the mixture.) →

2. For best results, cover the mixture with plastic wrap directly on the surface and chill overnight, or for at least 4 hours. The madeleines can be baked right away, but their trademark hump will be smaller.

3. When you are ready to bake, preheat the oven to 400 degrees F (200 degrees C). Grease the madeleine molds. Spoon small amounts of batter into the molds (1 tablespoon for standard madeleines). Bake the madeleines for 9 to 11 minutes, or until the edges are dark golden and the centers spring back when touched. Immediately tap out the madeleines onto a rack (they should release easily from the pan). Serve them warm or at room temperature. (If you don't have a madeleine pan, you could bake these in the bottoms of standard muffin tins. The look will be different, but the texture will be identical.)

4. Madeleines are best served warm the same day they are made. If you plan to hold them, as soon as they are cool, store them at room temperature in an airtight container for up to 2 days.

VARIATION

Orange, Rosemary, and Almond Brittle

To make Orange, Rosemary, and Almond Brittle, follow the recipe as given but use toasted, slivered almonds in place of hazelnuts.

Orange, Rosemary, and Hazelnut Brittle

A combination of Pacific Northwest rosemary and seasonal citrus complements the deep caramel flavor of this amber nut brittle. I favor using Oregon hazelnuts, but the act of peeling them can steer you quickly toward almonds, which work well in this recipe. (I usually use the slivered almonds.) Either way, this brittle's amber stained-glass appearance is spectacular packaged as holiday gifts.

Makes about 1 pound (½ kg)

1 cup (142 g) hazelnuts, toasted, peeled, and roughly chopped

2 teaspoons fresh rosemary, roughly chopped

1 teaspoon grated orange zest

1½ cups (300 g) sugar

½ cup (113 mL) mild honey or corn syrup

⅓ cup (76 mL) water

1. Spray a piece of foil, approximately 12 by 16 inches, with nonstick spray. Have a clean spatula ready.

2. Combine the nuts, rosemary, and orange zest in a bowl.

3. Combine the sugar, honey, and water in a medium pot and stir to combine. (Be sure that no dry sugar is left on the bottom of the pan.) Wash down any sugar crystals on the sides of the pot with a pastry brush dipped in cold water.

4. Cook the sugar mixture over high heat without stirring, until the sugar caramelizes to a medium-dark amber and reaches about 300 degrees F (150 degrees C). (It's OK to swirl the pot to even the amber color.) You can use a candy thermometer for a temperature, but it's also important to observe the color and aroma of →

the brittle to get a deep caramel without burning it. Working quickly, remove the sugar from the heat and stir in the nut mixture with the spatula.

5. Pour the brittle out onto the foil, and spread it quickly with the spatula as thin as possible. (Move the brittle to a cooler counter spot once it starts to set so that it cools quickly.)

6. When the brittle is completely cool, break it into pieces with gloved hands and store in an airtight container. The brittle will hold for 1 to 2 weeks and longer in a dry climate. Do not refrigerate the brittle. It's stunning packaged in cellophane bags and tied with festive ribbon.

DRIED FRUITS

In winter, we depend on preserved, dried fruit to get us through a season that is sparse on fresh, seasonal fruit. Dried fruit widens our options, leading us to fruits we might not otherwise have available, adds flavors and textures, and gives us new recipes that can't be made with fresh fruit. These are just a few of my favorite recipes using dried fruit.

Pick dried fruits that are moist. I'm particularly fond of dried figs, cherries, and California apricots (the tart ones). Overly dried fruits have lost their texture and some of their flavor. Fruits that have become too dry can be reconstituted by immersing them in boiled water for a few minutes, or simply by pouring boiling water over them in a colander. Let reconstituted fruits dry for thirty minutes or so before using.

Grandma's Dark Holiday Fruitcake

If you don't like fruitcake, you probably haven't tried this one. It's full of dates, raisins, and walnuts, with a small amount of other dried fruit. (No fake-colored fruit here.) This is the fruitcake my grandmother and then my mother made every Christmas. This recipe has been passed down and protected. My mother and I enjoy giving it away, but only to those friends we *know* will treasure it. The nutmeg-related spice, mace, is key here. (Mace is ground from the webbing that covers the outside of the nutmeg kernel.) Plan ahead to make this cake as it needs to cure for three to four weeks. My mother always made it a few days before Thanksgiving, and then she unveiled it on Christmas Eve.

Makes two 5-by-9-inch loaf cakes

For the fruit-and-nut overnight soak:
3 cups (454 g) raisins

3 cups (454 g) pitted dates, chopped into ½-inch pieces

3 cups (454 g) of other dried fruits, chopped into ½-inch pieces (any variety, such as Bing cherries, figs, pineapple, papaya, apricots, or currants)

2 cups (227 g) walnuts, chopped into ½-inch pieces

¼ cup (85 g) molasses

¼ cup (57 mL) red wine or port

For the fruitcake:
2 cups (260 g) all-purpose flour

1 teaspoon cinnamon

½ teaspoon ground cloves

½ teaspoon ground mace

½ teaspoon kosher salt

¼ teaspoon baking soda

1¼ cups (273 g) light or dark brown sugar

1 cup (2 sticks; 227 g) unsalted butter, room temperature, plus more for greasing

4 eggs, room temperature

½ cup (114 mL) red wine or port for post-bake soak, plus ¼ cup (57 mL) more for a second soak if desired

1. *To make the fruit-and-nut overnight soak,* combine all the ingredients for the soak in a very large bowl and toss well. Cover with plastic wrap or a lid, and let the fruit mixture sit for 24 hours at room temperature, stirring twice.

2. *To make the fruitcake,* preheat the oven to 275 degrees F (135 degrees C). Grease the loaf pans. Create two slings of parchment that extend over each long side of the pan and leave the ends uncovered. This allows you to remove the cake easily.

3. Sift the flour, cinnamon, cloves, mace, salt, and baking soda in a medium bowl and set aside. In the bowl of a stand mixer fitted with the paddle attachment, cream the sugar and butter on medium speed until light and fluffy, about 5 minutes, scraping once. Add the eggs one at a time until combined, scraping the bowl after each addition. Add the dry ingredients all at once and mix on low speed until the flour is just combined. Scrape the bowl one more time and mix for another 10 to 15 seconds.

4. Add the cake batter to the fruit mixture and stir together until combined. I use my hands to mix it because it is stiff.

5. Divide the cake batter among the pans, leveling the batter. Bake for 2 to 2 hours and 15 minutes, or until a toothpick inserted in the center comes out clean. Cool until slightly warm. Remove the cakes from the pans, and place each on cheesecloth folded two-ply. Poke holes in the top of the cakes with a skewer or thermometer probe, and pour the ½ cup (114 mL) wine or port over the two cakes. Wrap the cake in the cheesecloth and then tightly in foil, and place it in a cool, dry place for 3 to 4 weeks. (Optional: Resoak the cake with another ¼ cup (57 mL) port or wine after 1 week and rewrap.)

6. Once the cake has cured, keep it at room temperature for a few weeks or freeze it. I always keep it frozen because it's easy to cut. Just wait for it to come to room temperature before serving. This fruitcake will keep for several months in the freezer.

< Grandma's Dark Holiday Fruitcake ^ Apricot-Almond Scones

Apricot-Almond Scones

When someone asks me, "What is your favorite thing to make?", these scones usually are the first to come to mind. Not only are they incredibly tender and flavorful, but the best scone you will ever have will probably be the ones *you* bake. It seems difficult to find a good scone because most are dry and overworked, or soft, like a muffin. Getting the texture correct is similar to pie dough; it's all in the handling. A light touch along with quality ingredients will make a great scone every time. It's easy to change up the flavors in this scone by just replacing the fruit and nuts or substituting the extract for spices. I recommend doubling the recipe when you make these; bake one disk and freeze the other for up to one month.

1. Whisk together the flour, sugar, baking powder, and salt in a medium bowl.

2. In a small bowl, whisk together the cream, egg, and almond extract.

3. Cut the butter into ½-inch pieces, and add it to the flour mixture. Using your hands or a pastry cutter, work in the butter until it is slightly larger than pea size. (Work quickly to keep the butter cold.) Add the almonds and apricots to the dry mixture and toss to combine.

VARIATION

Lemon Balm Scones

Make the scones as written, omitting the apricots, almonds, and almond extract. Before adding the cream and egg, add 4 tablespoons chopped fresh lemon balm and 1 teaspoon grated lemon zest to the dry mixture. Continue with the scone recipe as written. Divide the scone disk into eight portions.

Makes 6 scones

1½ cups (195 g) all-purpose flour or pastry flour (see note, page 187)

¼ cup (25 g) sugar, plus more for finishing

2 teaspoons baking powder

¾ teaspoon kosher salt

½ cup (113 mL) heavy cream, plus more for finishing

1 egg

½ teaspoon almond extract

4 tablespoons unsalted butter, chilled

⅓ cup (34 g) sliced or slivered almonds, toasted

⅓ cup (43 g) dried apricots, diced small

4. Add the cream mixture all at once to the flour mixture and stir gently with a fork until the dough starts to hold together. (The dough will be ragged, shaggy, and moist. Don't worry about making it pretty.)

5. Turn the dough out onto a clean, dry surface. Gently gather the dough and pat it out to about 1 inch thick. Fold the dough in half, pat it out, and then fold in half again. (This will create some layers of dough and butter that will give the scone a nice flakiness.)

6. Shape the dough into a disk about 1 inch thick and 6 inches in diameter. Wrap the dough disk in plastic, and chill it in the freezer until firm, about 2 hours. (The dough can be frozen up to 1 month. Your scones will hold the best shape if they are mostly or completely frozen before you bake them.)

7. Preheat the oven to 375 degrees F (190 degrees C). Remove the disk from the freezer, and cut it into six triangles, like a pie. Dip the tops of the scones in heavy cream and then in sugar. (If you don't want to use the sugar topping, be sure to use the heavy cream for shine and color.) Place the scones on a parchment-lined baking sheet about 1 inch apart. Bake the scones for 25 to 30 minutes, until they are lightly brown and firm when touched. (If the scones start to brown too much on the bottom during baking, place a second cookie sheet under the first.)

8. Cool the scones on a rack, and store them in an airtight container at room temperature. These scones are best eaten the same day they are baked, but they'll still be tasty the second day.

Note: Pastry flour is a lower-protein flour, so it creates less gluten and a more tender product. If you don't have pastry flour on hand, you can combine two parts all-purpose flour with one part cake flour. I keep this combination handy in a separate container so I always have some ready to use. Whole wheat pastry flour is particularly tender and tasty in these scones.

ACKNOWLEDGMENTS

As I look back on the journey of writing this book, I realize it wouldn't have been possible to be inspired to write it without the dedicated farmers, orchardists, and fruit growers who toil in the fields so that we may enjoy grown-with-care, high-quality, flavorful fruit. I am especially in debt to the following local farms that grew the fruit that graces these pages: Bow Hill Blueberries, Collins Family Orchards, Hayton Farms, Heirloom Orchards, Martin Family Orchards, Sidhu Farms, Sky Harvest Produce, and Tonnemaker Hill Farm.

I waited a long time to write a cookbook (despite the urging from my students and colleagues), and I appreciate my family, friends, and others who gently encouraged me to write. It took me a while to get there. First, I would not be writing this without the influence of my mom and my late maternal grandmother, who were not only great bakers but great teachers. I acquired my love of fruit and pastry from both of you. Thank you to Susan Roxborough, Bryce de Flamand, and Jill Saginario at Sasquatch Books for your support, humor, advice, and guidance. I was immediately pleased to learn of and work with photographer Charity Burggraaf. Her gorgeous photos make the book that much better. Recipe testers are invaluable, and I owe a great deal to Lisa Dossett, Irene Eckhardt, Laura Eisenberg, Carla Fisher, Anne Forestieri, Betty Frost, Maria Galvao, Judy Green, Michelle Mancuso, Sheila Mitchell, Rebecca Morrison, Erin Ostrander, Sherrie Pierce, Diane Priebe, Ikue Tashima, Amy Tsang, and Denise Vaughn. A special thank-you to Becky Selengut, who showed me the ropes and helped set me on the path to writing my first book. Finally, thank you to all my students who have informally tested my recipes for the past ten years, giving feedback and praise and very constructive criticism. You have made me better.

INDEX

Note: Page numbers in italic refer to photographs.

A

alcohol-based desserts
- Brandied Apple Pie with Ginger Streusel, 138–39, *140*
- Burgundy-Poached Pears with Mulled-Wine Syrup, *151*, 152–53
- Pain Perdu with Assorted Berries and Grand Marnier, *85*, 86–87
- Poached Apricots in Sauternes, *104*, 105–06, *107*
- Port Sabayon with Fresh Raspberries, *54*, 55
- Rosemary Apples in Crepes with Rum Caramel Sauce, 134–35, *136*
- Smoky Sweet Cherries in Port with Bittersweet Chocolate, 41–42, *43*
- White Chocolate Mousse with Poached Sweet Cherries, 46–48, *49*

Almond Cake with Warm Citrus and Thyme, *172*, 173–74

apples
- about, *132*, 133
- Apple-Cranberry Tarte Tatin, *163*, 164–65
- Apple Upside-Down Cake with Spelt and Rye Flours, *141*, 142–43
- Brandied Apple Pie with Ginger Streusel, 138–39, *140*
- Cider-Braised Apples with Warm Winter Spices, 144–45, *145*
- Rosemary Apples in Crepes with Rum Caramel Sauce, 134–35, *136*

apricots
- about, *96*, 97
- Apricot-Almond Scones, *185*, 186–87
- Apricot and Walnut Streusel Bars, *101*, 102–03
- Grilled Apricots with Brown Butter and Maple-Tamari Glaze, 98–99, *100*
- Poached Apricots in Sauternes, *104*, 105–06, *107*

B

berries. See specific types of berries

Bittersweet Chocolate Tart with Blackberries and Basil, *88*, 89–90

blackberries and marionberries
- about, *80*, 81
- Bittersweet Chocolate Tart with Blackberries and Basil, *88*, 89–90
- Marionberry Crostata with Whole Grain Crust, 82–83, *84*
- Pain Perdu with Assorted Berries and Grand Marnier, *85*, 86–87
- Vanilla Bean Cake with Glazed Blackberries and Stone Fruits, 91–92, *93*

blueberries
- about, *64*, 65
- Blueberry and Lemon Curd Tart, *75*, 76–77
- Blueberry-Cinnamon Crepes, 70–71, *71*, *74*
- Classic Lattice Blueberry Pie, *66*, 67–69
- Poached Blueberries with Vanilla Bean and Anise, *78*, 79

Brandied Apple Pie with Ginger Streusel, 138–39, *140*

bread-based desserts
- Pain Perdu with Assorted Berries and Grand Marnier, *85*, 86–87
- Roasted Peach Bread Pudding, 116–17, *119*

brittle
- Orange, Rosemary, and Almond Brittle, 177–78
- Orange, Rosemary, and Hazelnut Brittle, 177–78, *179*

Burgundy-Poached Pears with Mulled-Wine Syrup, *151*, 152–53

C

cake
- Almond Cake with Warm Citrus and Thyme, *172*, 173–74
- Apple Upside-Down Cake with Spelt and Rye Flours, *141*, 142–43
- Classic White Cake, 94
- Cranberry and Bay Upside-Down Cake, 166–67, *167*
- Dark Chocolate Cake with Raspberry-Orange Compote, *61*, 62–63
- Grandma's Dark Holiday Fruitcake, 182–83, *184*
- Strawberry-Lime Layer Cake, *22*, 23–25

Candied Lime Zest, Quick, 26, *27*

cheesecake
- Fromage Blanc Cheesecake with Brown-Butter-Braised Plums, 128–30, *131*

cherries
- about, *36*, 37
- Cherry Hand Pies, *38*, 39–40

Smoky Sweet Cherries in Port with Bittersweet Chocolate, 41–42, *43*
Sour Cherry Compote, *44*, 45
White Chocolate Mousse with Poached Sweet Cherries, 46–48, *49*

chocolate
Bittersweet Chocolate Tart with Blackberries and Basil, *88*, 89–90
Dark Chocolate Cake with Raspberry-Orange Compote, *61*, 62–63
Smoky Sweet Cherries in Port with Bittersweet Chocolate, 41–42, *43*
White Chocolate Mousse with Poached Sweet Cherries, 46–48, *49*

Cider-Braised Apples with Warm Winter Spices, 144–45, *145*

citrus
about, *168*, 169
Almond Cake with Warm Citrus and Thyme, *172*, 173–74
Blueberry and Lemon Curd Tart, *75*, 76–77
Dark Chocolate Cake with Raspberry-Orange Compote, *61*, 62–63
Meyer Lemon Buttermilk Sherbet, 170, *171*
Orange, Rosemary, and Almond Brittle, 177–78
Orange, Rosemary, and Hazelnut Brittle, 177–78, *179*
Orange and Rye Madeleines, 175–76, *176*
Quick Candied Lime Zest, 26, *27*

coconut
Strawberry-Coconut Sorbet, 28–29, *30*

compote
Dark Chocolate Cake with Raspberry-Orange Compote, *61*, 62–63
Sour Cherry Compote, *44*, 45

cranberries
about, *158*, 159
Apple-Cranberry Tarte Tatin, *163*, 164–65
Cranberry and Bay Upside-Down Cake, 166–67, *167*
Mom's Cranberry-Walnut Pie, 160–61, *162*

Crème Anglaise, Ginger, 118, *119*

crepes
Blueberry-Cinnamon Crepes, 70–71, *71*, *74*
Crepes, 72–73
Rosemary Apples in Crepes with Rum Caramel Sauce, 134–35, *136*

Crisp, Plum-Lavender, *126*, 127

D

Dark Chocolate Cake with Raspberry-Orange Compote, *61*, 62–63
Dark Chocolate Mousse, 46

dough
how to roll out pastry dough, 13
Flaky Pie Dough, 12
Sweet Tart Dough (Pâte Sucrée), 58–59

dried fruits
about, *180*, 181
Apricot-Almond Scones, *185*, 186–87
Grandma's Dark Holiday Fruitcake, 182–83, *184*

E

equipment
bench knife, xi
bowl scraper, xi
Microplane zester, xi
pastry wheel, xi
spatulas and whisks, xi

F

figs
Pear and Fig Pie, 148–49, *150*

Flaky Pie Dough, 12
Fromage Blanc Cheesecake with Brown-Butter-Braised Plums, 128–30, *131*
frozen desserts. *See* sorbet and other frozen desserts
fruit. *See* specific types of fruit
dried fruits, *180*, 181–87
using fresh *vs.* frozen, ix
Fruitcake, Grandma's Dark Holiday, 182–83, *184*
Fruit Sorbet, 6

G

ginger
Brandied Apple Pie with Ginger Streusel, 138–39, *140*
Ginger Crème Anglaise, 118, *119*
Rhubarb-Ginger Sorbet, 6, 7

Grandma's Dark Holiday Fruitcake, 182–83, *184*
Grilled Apricots with Brown Butter and Maple-Tamari Glaze, 98–99, *100*

H

Hand Pies, Cherry, *38*, 39–40

herbs
Almond Cake with Warm Citrus and Thyme, *172*, 173–74
Cranberry and Bay Upside-Down Cake, 166–67, *167*
Orange, Rosemary, and Almond Brittle, 177–78
Orange, Rosemary, and Hazelnut Brittle, 177–78, *179*
Peaches and Nectarines Sautéed with Thyme, 110, *111–12*
Plum-Lavender Crisp, *126*, 127
Raspberry and Lemon Balm Shortcake, 52, *53*
Rosemary Apples in Crepes with Rum Caramel Sauce, 134–35, *136*

I

ice cream. *See* sorbet and other frozen desserts
ingredients
- eggs and dairy, ix
- flour, ix
- fresh *vs.* frozen fruit, ix
- sugar, x
- vanilla beans, x

L

Lattice Blueberry Pie, Classic, *66*, 67–69
layered desserts
- Rhubarb Fool, 14–15, *15*

lemon(s)
- Blueberry and Lemon Curd Tart, *75*, 76–77
- Meyer Lemon Buttermilk Sherbet, 170, *171*

Lemon Balm Scones, 186
lime(s)
- Quick Candied Lime Zest, 26, *27*

M

Madeleines, Orange and Rye, 175–76, *176*
Maple and Pear Panna Cotta, 154–55, *155*
Marionberry Crostata with Whole Grain Crust, 82–83, *84*
Meyer Lemon Buttermilk Sherbet, 170, *171*
Mom's Cranberry-Walnut Pie, 160–61, *162*
mousse
- Dark Chocolate Mousse, 46
- Peach Mousse, *113*, 114–15
- White Chocolate Mousse with Poached Sweet Cherries, 46–48, *49*

N

nuts
- Apricot-Almond Scones, *185*, 186–87
- Apricot and Walnut Streusel Bars, *101*, 102–03
- Grandma's Dark Holiday Fruitcake, 182–83, *184*
- Mom's Cranberry-Walnut Pie, 160–61, *162*
- Orange, Rosemary, and Almond Brittle, 177–78
- Orange, Rosemary, and Hazelnut Brittle, 177–78, *179*

O

orange(s)
- Dark Chocolate Cake with Raspberry-Orange Compote, *61*, 62–63
- Orange, Rosemary, and Almond Brittle, 177–78
- Orange, Rosemary, and Hazelnut Brittle, 177–78, *179*
- Orange, Rosemary, and Almond Brittle, 177
- Orange and Rye Madeleines, 175–76, *176*

P

Pain Perdu with Assorted Berries and Grand Marnier, *85*, 86–87
Panna Cotta, Maple and Pear, 154–55, *155*
Pâte Sucrée (Sweet Tart Dough), 58–59
Pavlova, Strawberry, *18*, 19–20, *21*
Pavlova, Family-Style, 19
peaches and nectarines
- about, *108*, 109
- Peaches and Nectarines Sautéed with Thyme, 110, *111–12*
- Peach Mousse, *113*, 114–15
- Roasted Peach Bread Pudding, 116–17, *119*

pears
- about, *146*, 147
- Burgundy-Poached Pears with Mulled-Wine Syrup, *151*, 152–53
- Maple and Pear Panna Cotta, 154–55, *155*
- Pear and Fig Pie, 148–49, *150*

pie. *See also* tarts
- fruit filling thickener suggestions, 69
- how to roll out pastry dough, 13
- Brandied Apple Pie with Ginger Streusel, 138–39, *140*
- Cherry Hand Pies, *38*, 39–40
- Classic Lattice Blueberry Pie, *66*, 67–69
- Flaky Pie Dough, 12
- Marionberry Crostata with Whole Grain Crust, 82–83, *84*
- Mom's Cranberry-Walnut Pie, 160–61, *162*
- Pear and Fig Pie, 148–49, *150*
- Straight Rhubarb Pie, 9–11, *10*
- Whole Wheat Pie Dough, 12

plums
- about, *122*, 123
- Fromage Blanc Cheesecake with Brown-Butter-Braised Plums, 128–30, *131*
- Plum-Lavender Crisp, *126*, 127
- Simple Stewed Plums and Pluots, 124, *125*

poached fruit
- Burgundy-Poached Pears with Mulled-Wine Syrup, *151*, 152–53
- *Cider-Braised Apples with Warm Winter Spices, 144–45, 145*
- Poached Apricots in Sauternes, *104*, 105–06, *107*
- Poached Blueberries with Vanilla Bean and Anise, *78*, 79
- White Chocolate Mousse with Poached Sweet Cherries, 46–48, *49*

Port Sabayon with Fresh Raspberries, *54*, 55

Q

Quick Candied Lime Zest, 26, *27*

R

raspberries
- about, *50*, 51
- Dark Chocolate Cake with Raspberry-Orange Compote, *61*, 62–63
- Pain Perdu with Assorted Berries and Grand Marnier, *85*, 86–87
- Port Sabayon with Fresh Raspberries, *54*, 55
- Raspberry and Lemon Balm Shortcake, 52, *53*
- Raspberry Custard Tart, 56–57, *60*

rhubarb
- about, *2*, 3
- Infused Rhubarb, 14
- Rhubarb Fool, 14–15, *15*
- Rhubarb-Ginger Sorbet, 6, *7*
- Straight Rhubarb Pie, 9–11, *10*
- Vanilla Roasted Rhubarb, *4*, 5

Roasted Peach Bread Pudding, 116–17, *119*
Roasted Strawberry Ice Cream, *31*, 32–33
Rosemary Apples in Crepes with Rum Caramel Sauce, 134–35, *136*
Rum Caramel Sauce, *136*, 137

S

sauce(s)
- Ginger Crème Anglaise, 118, *119*
- Rum Caramel Sauce, *136*, 137

scones and shortcake
- Apricot-Almond Scones, *185*, 186–87
- Lemon Balm Scones, 186
- Raspberry and Lemon Balm Shortcake, 52, *53*

Smoky Sweet Cherries in Port with Bittersweet Chocolate, 41–42, *43*
sorbet and other frozen desserts
- how to freeze, 8
- using sugar in, 8
- Fruit Sorbet, 6
- Meyer Lemon Buttermilk Sherbet, 170, *171*
- Rhubarb-Ginger Sorbet, 6, *7*
- Roasted Strawberry Ice Cream, *31*, 32–33
- Strawberry-Coconut Sorbet, 28–29, *30*

Sour Cherry Compote, *44*, 45
Stewed Plums and Pluots, Simple, 124, *125*
stone fruit
- Apricot and Walnut Streusel Bars, *101*, 102–03
- Fromage Blanc Cheesecake with Brown-Butter-Braised Plums, 128–30, *131*
- Grilled Apricots with Brown Butter and Maple-Tamari Glaze, 98–99, *100*
- Peaches and Nectarines Sautéed with Thyme, 110, *111–12*
- Peach Mousse, *113*, 114–15
- Plum-Lavender Crisp, *126*, 127
- Poached Apricots in Sauternes, *104*, 105–06, *107*
- Roasted Peach Bread Pudding, 116–17, *119*
- Simple Stewed Plums and Pluots, 124, *125*
- Vanilla Bean Cake with Glazed Blackberries and Stone Fruits, 91–92, *93*

Straight Rhubarb Pie, 9–11, *10*
strawberries
- about, *16*, 17
- Roasted Strawberry Ice Cream, *31*, 32–33
- Strawberry-Coconut Sorbet, 28–29, *30*
- Strawberry-Lime Layer Cake, *22*, 23–25
- Strawberry Pavlova, *18*, 19–20, *21*
- Vanilla Roasted Strawberries, 5

Sweet Tart Dough (Pâte Sucrée), 58–59

T

tarts. *See also* pie
- Apple-Cranberry Tarte Tatin, *163*, 164–65
- Bittersweet Chocolate Tart with Blackberries and Basil, *88*, 89–90
- Blueberry and Lemon Curd Tart, *75*, 76–77
- Raspberry Custard Tart, 56–57, *60*

V

vanilla
- Poached Blueberries with Vanilla Bean and Anise, *78*, 79
- Vanilla Bean Cake with Glazed Blackberries and Stone Fruits, 91–92, *93*
- Vanilla Roasted Rhubarb, *4*, 5
- Vanilla Roasted Strawberries, 5

W

White Cake, Classic, 94
White Chocolate Mousse with Poached Sweet Cherries, 46–48, *49*
Whole Wheat Pie Dough, 12

ABOUT THE AUTHOR

Laurie Pfalzer encourages lively discussions with students about baking, pastry techniques, and seasonal produce. She is particularly passionate about teaching not only the how, but also the why of baking and pastry. Formerly the pastry chef at Salish Lodge and Spa, Laurie also worked under Jeffrey Hamelman at the King Arthur Flour Bakery in Norwich, Vermont. Laurie graduated with honors from the Culinary Institute of America in Hyde Park, New York. She lives in the Seattle area with her dog, Tucker, where she teaches baking and pastry to home cooks and consults for independent baking businesses.

Printed in China

SASQUATCH BOOKS with colophon is a registered trademark of Penguin Random House LLC

24 23 22 21 20   9 8 7 6 5 4 3 2 1

Editor: Susan Roxborough
Production editor: Jill Saginario
Photography: Charity Burggraaf
Designer: Bryce de Flamand

Library of Congress Cataloging-in-Publication Data

Names: Pfalzer, Laurie, author.
Title: Simple fruit / by Laurie Pfalzer.
Description: Seattle, WA : Sasquatch Books, [2020]
Identifiers: LCCN 2019003134 | ISBN 9781632172372 (hard cover)
Subjects: LCSH: Desserts. | Cooking (Fruit) | Berries. | Fruit. | LCGFT: Cookbooks.
Classification: LCC TX773 .P467 2020 | DDC 641.6/4—dc23
LC record available at https://lccn.loc.gov/2019003134

ISBN: 978-1-63217-237-2

Sasquatch Books
1904 Third Avenue, Suite 710
Seattle, WA 98101

SasquatchBooks.com